Wind Power Projects

Wind power has developed rapidly in terms of the number of new wind power plants now installed in more than one hundred countries around the world. This renewable energy source has become competitive, and to be able to combat climate change much more has to be installed in coming years. This also makes it necessary for policy makers, NGOs, research scientists, industry and the general public to have a basic understanding of wind power.

The majority of texts on wind power are written primarily for engineers or policy analysts. This book specifically targets those interested in, or planning to develop wind power projects. It can be understood by both specialists and non-specialists interested in wind power project development.

Having outlined the background of wind power and its development, explained wind resources and technology, the author explores the interactions between wind power and society and the role of wind power in the electric power system. Finally the main aspects of project development, including siting, economics and legislation, are explained.

This book will be an essential reference, or even a manual, for professionals developing new sites and for government officials and consultants involved in the planning or permission process. It can also be used as a textbook on wind power at schools and universities.

Tore Wizelius is a writer who has authored eight books on wind power. He was a lecturer on wind power at Gotland University, Sweden, from 1998 to 2008 and is now working as a wind power developer in Sweden.

Wind Power Projects

Theory and practice

Tore Wizelius

First published 2015
by Routledge
2 Park Square, Milton Park, Abingdon, Oxon OX14 4RN

and by Routledge
711 Third Avenue, New York, NY 10017

Routledge is an imprint of the Taylor & Francis Group, an informa business

British Library Cataloguing in Publication Data
A catalogue record for this book is available from the British Library

Library of Congress Cataloging in Publication Data
Wizelius, Tore, 1947–
Wind power projects : theory and practice / Tore Wizelius.
pages cm
Includes bibliographical references.
1. Wind power. I. Title.
TJ825.W593 2015
333.9´2–dc23 2014038962

ISBN: 978-1-138-78045-3 (hbk)
ISBN: 978-1-315-77071-0 (ebk)

Typeset in Sabon
by HWA Text and Data Management, London

Printed and bound in the United States of America by Publishers Graphics, LLC on sustainably sourced paper.

Contents

Figures

Tables

Boxes

1 Development of wind power

Wind power has grown rapidly since the end of the 1970s. In these few decades wind power has developed from an alternative energy source to a new fast-growing industry, which manufactures wind turbines that produce power at competitive prices.

Wind turbines take kinetic energy in the wind and transform it into mechanical work (in water pumps and windmills) or electric power (in modern wind turbines). The wind is a renewable energy source; the wind is set in motion by the differences in temperature and air pressure created by the sun's radiation. Wind turbines do not require any fuel to be transported – a hazard to the environment – don't create any air pollution and don't leave any hazardous waste behind. Wind turbines produce *clean* energy.

The sun, the wind and running water are all renewable energy sources, in contrast to coal, oil and gas, which depend on fossil deposits from mines, oil- or gas-fields. In most countries, hydropower has already been fully developed. The technology to use direct solar radiation with solar collectors and photovoltaic (PV) panels has had a commercial breakthrough, but development is still about 10 years behind. The new renewable energy source that has the most successful development so far is wind power, but during 2014 more PV-cells are expected to be installed than wind power.

Modern wind turbines are efficient, reliable and produce power at reasonable cost. The wind power industry is growing very fast; the leading companies have, in the first years in this decade, increased their turnover by 30 to 40 per cent per year. Simultaneously the cost per kWh of electicity produced has been lower for each new generation of wind turbines that has been introduced on the market.

From the early 1980s the size of wind turbines has doubled approximately every four years. The current generation of wind turbines has a rated power of 3–6 MW, rotor diameters of 120 metres or even more, and towers up to 140 metres in height. New prototypes are even larger and mainly intended for the offshore market.

The technology in the wind turbines has developed in several ways. The control systems have become cheaper and more advanced. New profiles for the rotor blades can extract more power from the wind, and new electronic power equipment makes it possible to use variable speeds and to optimize the capacity of the turbines.

In the same way as the wind turbines have grown in size, the installations have become larger and larger. In the early days of wind power development, the turbines were installed one at the time, often next to a farm. After a few years they were installed in groups of two to five turbines. Today, large wind power plants are built, on land and offshore, with the same capacity as conventional power plants. The largest wind power plants consist of hundreds of wind turbines and such large projects have now been developed in many parts of the world.

A challenge with the wind as an energy source is that the wind always varies. When the wind slows down or stops, power has to be produced by other power plants. This could lead to the conclusion that it will be necessary to have backup capacity with other power plants with the same capacity as the wind power connected to the power system. If this were true, wind power would be very expensive. Since wind power only constitutes one part in a large power system, this is, however, not necessary at all. A moderate proportion of wind power in a system does not need any backup capacity at all, since it already exists in the power system. In Scandinavia the power companies can simply save water in hydropower dams when the wind is blowing, and use this saved hydropower when the wind slows down.

In a power system, the power consumption varies continually, during the day as well as seasonally. Every power system has a regulating capacity to adapt power production to actual power consumption. This can be used to adapt the system to the variations in the wind – and the output of wind turbines – as well. When wind power penetration (that is the proportion of electric power produced by wind in a power system) increases to 10–20 per cent, it may be necessary to regulate the wind power as well, by reducing the power from wind turbines in situations with low loads (consumption) and high production, or by keeping a power reserve that can be used to balance power production with consumption at short notice. Few countries have however reached such penetration levels yet.

During the development from single small wind turbines connected to farms, to large wind power plants with the capacity of utility-scale power plants, wind power has become more competitive. The power produced by wind turbines has become cheaper. Today the cost of power produced by wind turbines (in places with good wind conditions) is competitive with the cost of power produced with oil, coal, gas or nuclear fuel in *new* power plants. In this decade wind power has become one of the cheapest energy sources available.

During the last few years, the market in China has expanded most and at the start of 2015 China had most installed capacity, followed by the USA where large wind power plants are installed on the plains of the Midwest and on the west coast. In Europe, Germany passed pioneering Denmark in 1994 with respect to installed wind power capacity, followed by Spain where

Table 1.1 Global wind power 2014/15 (MW)

Country	*Installed 2014*	*Total 2014/15*
China	23,351	114,763
USA	4,854	65,879
Germany	5,279	39,165
Spain	28	22,987
India	2,315	22,465
UK	1,736	12,440
Canada	1,871	9,694
France	1,042	9,285
Italy	108	8,663
Sweden	1,050	5,425
Denmark	67	4,845
Others	7,203	43,883
Total	51,477	369,553

Sources: GWEC, 2015; EWEA, 2015

a massive development of several thousand megawatts took place during the first decade of the twenty-first century (see Table 1.1 and Figure 1.4). There is still an immense potential for market growth in countries where development recently has taken off, like Australia, Brazil, Ireland, Canada and Poland, just to mention a few.

There are also ambitious plans to develop wind power plants offshore. Several offshore wind farms are already installed, in Denmark, Great Britain, the Netherlands and Sweden. Denmark has decided that wind power shall produce 50 per cent of the electric power in the country by 2030, and by the development of large offshore wind power plants this ambitious target is already within reach. Great Britain has also implemented an ambitious plan for offshore development.

In 2014 a further 51,500 MW of wind power was connected to the electric power grids around the world, and total installed capacity increased to 370 GW, an increase by 16 per cent from 2014. These 370 GW wind power produce around 755 TWh a year, which supplies some 170 million households with electricity. In Denmark wind power supplied 39 per cent of the total electric power consumption in 2014.

If this power had been generated by coal-fired power plants, some 255 million tons of coal would have been used, 255,000 railway wagons or 12,000,000 road transport lorries and some 530 million tons of carbon dioxide emitted.

Wind power has developed very quickly during the last 30 years. In the late 1970s and early 1980s wind turbines were small, manufactured in

Box 1.1 Wind power statistics

To indicate how much wind power there is in a country, the total installed capacity is used as a measure. Every wind turbine has a rated power (maximum power) that can vary from a few hundred Watts to 5,000 kW (5 MW). The number of turbines does not give any information on how much wind power they can produce. How much a wind turbine can produce depends not only on its rated power, but also on the wind conditions. To get an indication of how much a certain amount of installed (rated) power will produce, this simple rule of thumb can be used; 1 MW wind power produces 2 GWh/a (gigawatts per annum) on land and 3 GWh/a offshore. For the multi-MW turbines introduced during the last years these rules of thumb must be updated to: 1 MW produces 2.5 GWh/a on land and 4 GWh/a offshore.

1 TWh (terawatt hours) = 1,000 GWh (gigawatt hours)
1 GWh = 1,000 MWh (megawatt hours)
1 MWh = 1,000 kWh (kilowatt hours)
1 kWh = 1,000 Wh (watt hours)

International information on wind turbine installations is available at www.wwindea.org, www.ewea.org, www.gwec.net and on www.ieawind.org

smithies and small workshops and installed next to farms. Thirty years later wind power plants on land and offshore are integrated parts of the dominant energy system, and are manufactured by industrial giants such as General Electric in the United States, Siemens in Germany and their counterparts in India and China. In 2014, around 5 per cent of all electric power in the world was produced by wind turbines (WWEA, 2015) a share that will grow for many years ahead.

The power in the wind has been utilized by sailing ships and for different kinds of windmills for thousands of years (Hills, 1996). Wind turbines for electric power production were developed in Denmark by Paul laCour, and the first commercial grid-connected wind turbine started operating in Askov, Denmark, in 1902. From the 1920s battery-charging wind turbines were in common use in the countryside in the United States until the power grid was extended to these areas some decades later. This kind of small wind turbines were quite common also in Europe.

In the 1950s a wind turbine which delivered power to the AC grid was developed, the so-called Gedser-möllan, which gave valuable experience to Danish researchers and electrical engineers when it was in operation in the 1950s, but the abundance of fossil fuels that flooded the market at that time made it difficult for wind power to be competitive.

The modern wind power industry was developed in Denmark, starting in the late 1970s and it has been growing ever since. This was the start of the Danish success story in the wind power industry, which some 40 years later has spread across the globe.

Ever-growing wind turbines

Commercial wind turbines for grid connection, which are the focus of this book, were rather small when development started in the late 1970s and early 1980s. At that time the typical wind turbines had hub heights of about 20 metres, and on these lattice towers there was a nacelle with a 20–40 kW generator, a rotor with a diameter of around 15 metres and a swept area of 175 m^2 (see Figure 1.1).

In 2010 the largest wind turbines had hub heights of up to 160 metres, nominal power up to 7.5 MW, and rotor diameters close to 130 metres (see Figure 1.2). The industry is now developing wind turbines with a nominal power of 10–15 MW.

Beginning in the late 1970s wind turbine sizes have doubled on average every three to four years, if size is measured as the most common nominal power of the turbines during a specific year. When nominal power is increased, hub heights and rotor sizes increase as well (see Figure 1.3).

Figure 1.1 Typical wind turbine from early 1980s. A Vestas V15 with 35 kW nominal power and 18 m hub height installed at Lövsta Agricultural School on Gotland (photo: Tore Wizelius)

Figure 1.2 Gamesa G128 with 4.5 MW nominal power, 120 metre hub height and 128 metre rotor diameter is a new large wind turbine from the Spanish manufacturer Gamesa (photo: Tore Wizelius)

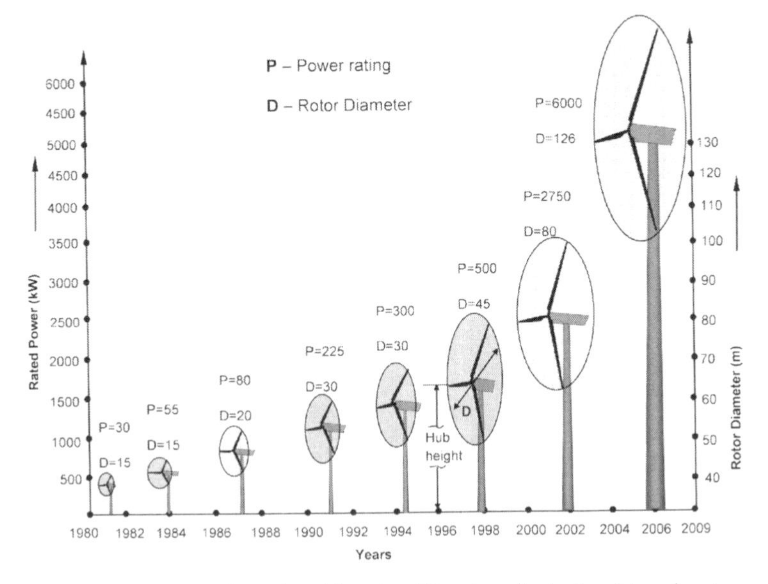

Figure 1.3 Development of wind turbine size. The size of wind turbines has increased quickly, with a doubling each three to four years (Earnest & Wizelius, 2011)

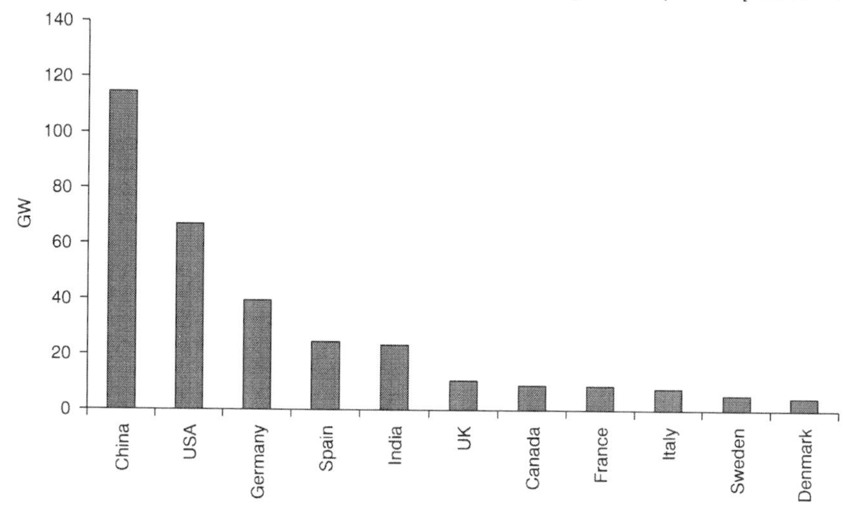

Figure 1.4 Installed capacity (GW) by country in 2014/15 (source: author's elaboration, based on data from GWEC 2015)

Early starters and late boomers

Denmark, a rather small country in northern Europe, has been at the forefront of the technical development of wind turbines as well as the actual installation of wind power connected to the power grid. An historical overview of the installed capacity of wind power in the electric power system shows that Denmark was in the lead from the 1970s and up to the early 1990s, except during the short wind power boom in California in the 1980s.

In 1994 Germany overtook Denmark and kept the lead position for a decade, with Spain in second place in the early years of the twenty-first century, the US taking the lead in 2007 and China, with an annual growth rate of more than 100 percent per annum for several consecutive years, taking over the lead position in 2010 (see Figure 1.4).

It may not be fair to compare a small country like Denmark with giants like India, China and the United States; a better measure for comparison is the wind power *penetration*, which means the proportion of electricity supplied by wind power to the national electric power system. In this respect Denmark was still in the lead in 2014. China and the United States, with most installed capacity, are below Sweden on this list (see Table 1.2).

Other measures are wind power capacity per capita and wind power capacity per land area in different countries (see Figures 1.5 and 1.6).

Table 1.2 Wind power penetration, 2013

Country	*Penetration, %*
Denmark	33
Portugal	17
Spain	16
Ireland	15
Germany	12
....	
Sweden	7

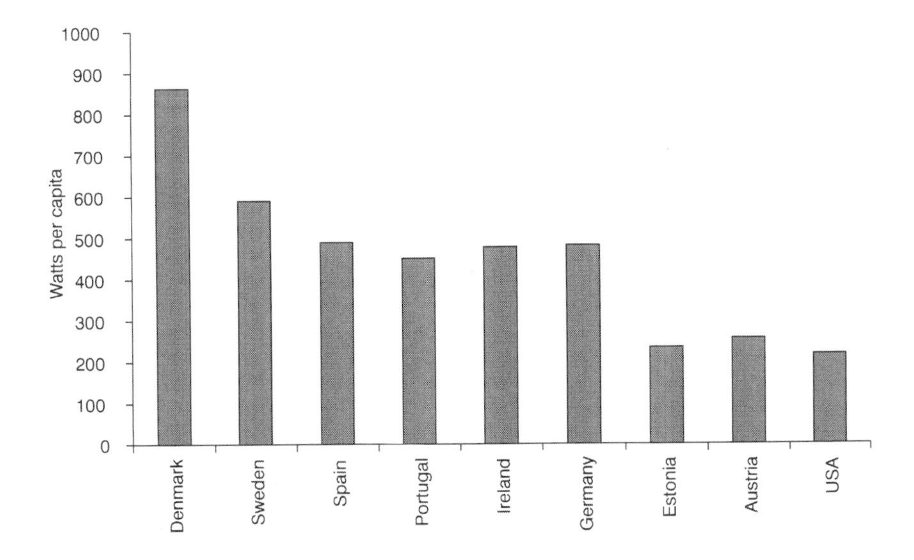

Figure 1.5 Wind power capacity per capita, 2014/15. In relation to the population Denmark is still far ahead of other countries; by this measure Sweden is in second position with the same level as Spain (source: author's elaboration, based on data from GWEC 2015, EWEA 2015)

Wind power industry

The development of the wind power industry started when the oil crises during the 1970s created panic in the industrialized world, which was enhanced in the 1980s by the nuclear reactor accidents in Harrisburg and Chernobyl. Politicians, as well as the public, were looking for new energy sources. The option was to develop local renewable energy sources; solar energy, biomass fuels, hydropower and wind power.

In the 1990s the focus was moved to the ever-increasing emissions of carbon dioxide and other greenhouse gases that threaten to change the

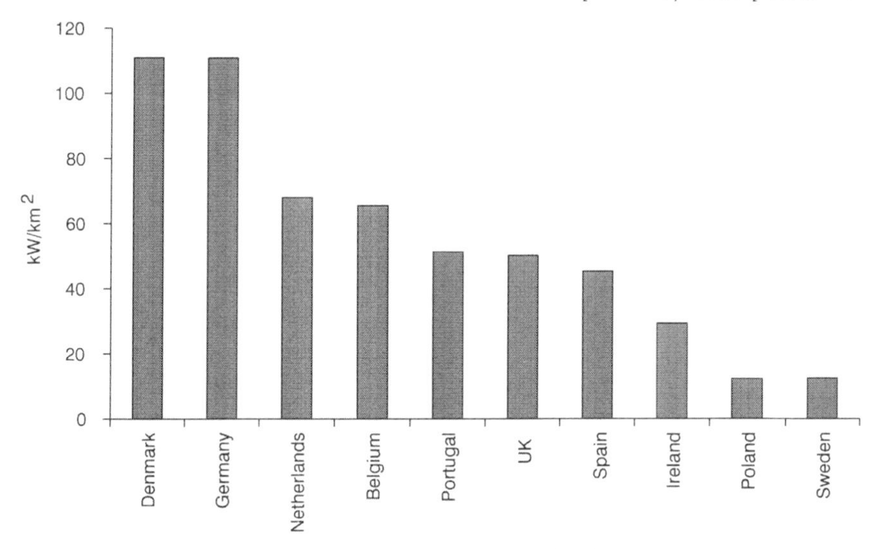

Figure 1.6 Wind power capacity per land area (kW/km²), 2014/15. In relation to land area, Germany has caught up with Denmark and shares the lead, and by this measure Sweden is far behind. This shows that there is plenty of space for more wind power development in most countries of the world (source: author's elaboration, based on data from GWEC 2015, EWEA 2015)

Figure 1.7 The end for the prototype Näsudden in Sweden. In June 2008 the Swedish 3MW prototype at Näsudden was demolished to make space for a modern wind turbine. An industrial heritage site was destroyed; in a hundred years this could have been a monument over the effort to avoid the global climate change (photo: Ivo Palu)

global climate. The burning of fossil fuels for power production was the main source of these emissions; energy policy and environment policy turned out to have a close connection. Since the UN Environment Conference in Rio de Janeiro in 1992 the necessity to change to renewable energy sources has been given even more weight.

Many governments then decided to invest money in research and development and entrusted the task to develop wind power to large power companies and the aircraft industry, since they should have the competence and the resources to succeed.

Sweden, Germany, the United Kingdom, the Netherlands, the United States and many other countries started national wind power research programmes. The aim was to develop large wind turbines, with a nominal power of several megawatts, which were considered necessary for wind power to make a significant contribution to the power production on a national level.

During the 1980s several very large R&D wind turbines were erected in different countries. In Sweden large prototypes with a nominal power of 2–3 MW, Näsudden I and Maglarp, were built. Wind turbines of this size were built in the United States and Germany too.

These behemoths, as the author Paul Gipe named them in his book *Wind Energy Comes of Age* (Gipe, 1995) weren't succesful, but they provided experience that has been useful for the wind power industry.

The Danish success story

The Danes choose a completely different strategy: to create a market for wind turbines. Generous investment grants were introduced, and the state guaranteed a good and reliable price for the wind-generated electric power. Investors were assured that they would get their money back. All political parties in the Danish parliament supported this policy. There was a political consensus.

In Denmark, completely different kinds of companies started to manufacture wind turbines; small and medium-sized enterprises, workshops and smithies that produced ploughs and other farming equipment. They manufactured small wind turbines, 20–100 kW for the farm market (see Figure 1.1).

Farmers want machines that are reliable and robust, stand all kinds of weather conditions and if they break down, it should be possible to repair them in the farm workshop. Danish companies manufactured wind turbines according to these principles. They did not opt for high-tech, but for simplicity and durability. The first models were 15–20 metres tall with a nominal power of 20–30 kW, which every year grew a little. These companies had a large and loyal group of customers, a solid market, and wind turbines became just another machine among their products.

Many farmers bought their own wind turbines, even though it was a new technology with considerable financial risk. Members of the public, worried about the environmental crisis, formed wind power cooperatives.

The wind power research centre at Risoe tested and certified the wind turbines. Manufacturers got feedback from the researchers, who had many good ideas about how the wind turbines could be further developed. Simultaneously the number of wind turbines on-line grew, and researchers and manufacturers got valuable feedback from the practical operation of hundreds of wind turbines.

In the early 1980s the wind turbines had a nominal power of about 20 kW, by the year 2000 the size was approaching 2,000 kW (2 MW) and in 20 years the size of the wind turbines increased by a factor 100. The strategy to create a market for commercial wind power by investment grants and other subsidies in the 1980s, turned out to be a much more successful strategy than large investments in high-tech research and development projects.

Germany and Spain catch up with the Danes

Germany introduced a 100 MW programme for wind power development that soon was extended to a 250 MW programme. The Danes got a new fast-growing export market. However, a large share of the support went to German manufacturers who started to compete with the Danes. In Germany and Denmark, economic support to wind power was not seen as only a part of the energy and environment policy, but also as part of the industrial policy; support will make the manufacturing companies grow, so that new jobs are created and economic growth will increase.

From the year 2000, Spain has had a remarkable growth in the wind power sector, and the Spanish power companies play an active role in this development. Also in Spain the political will to develop wind power on a large scale is not only a part of energy policy, but of industrial and regional policy as well. Many old shipyards in the north-western Atlantic coast have been taken over by wind power companies and a lot of new jobs have been created in this once depressed region of Spain. A domestic wind turbine manufacturing industry was developed as well.

The Asian pioneer India and the boom in China

India has played a very prominent role as a pioneer for wind power in Asia. In the 1990s several European manufacturers established their own factories or joint ventures, and domestic companies also began to manufacture wind turbines. Today the most successful Indian company, Suzlon, competes with European manufacturers on the global market (Earnest & Wizelius, 2011).

In China a similar development started a decade later. China was a late starter in the wind power business, compared with Europe and its Asian neighbour India. In 1995 only 38 MW were installed in a few small demonstration projects, and with modest annual development in the range of 50–100 MW up to 2003. In 2004, 197 MW were installed and the wind power capacity in 2004/05 totalled just 764 MW.

In the following years a tremendous growth started, with growth rates in the range of 100 per cent for several consecutive years; 503 MW (2004), 1,337 MW (2005), 3,304 (2006), 6,246 (2007), 12,209 (2008), 25,810 (2009), adding up to a total of 115,000 MW in 2013/14, moving China up to number one in the world. This rapid growth seems set to continue for many more years. Wind power has become a success story for China.

This rapid market growth had so far mainly been concentrated to just a few countries: Denmark, Germany, Spain, US, India and China. In 2007–2008 the development took off in countries like Canada, Australia, Brazil, France and Portugal. Today wind power is utilized – at varying scales – on all continents and in 100 countries and regions in the world (WWEA, 2013). Countries where development has so far hardly taken off will have rapid market growth in coming years.

Commercial wind turbines are now bigger than the prototype behemoths that were built with huge R&D funds in the 1980s. There is however one important difference; the commercial wind turbines of today are competitive and reliable: they work. And for each increase in size, the cost per kWh of electricity produced has decreased.

During the same period the projects have also grown larger. From installations with single wind turbines at farms in the 1980s, groups of 3–10 wind turbines and even larger wind farms in the 1990s, focus in Europe from 2000 and onwards has been on the development of large offshore wind farms, with big power companies as investors. Today wind power is a major industrial branch in many countries and has become big business.

Supply and demand

From 2005 up to 2008, demand for wind turbines was much larger than the supply. Developers who ordered wind turbines for their projects had to wait for years to get them delivered. It was also hard to place orders for single wind turbines, big orders had priority. And with a sellers' market, prices for wind turbines increased. All leading manufacturers, as well as subcontractors which manufacture components, were expanding their factories as fast as they could.

After a period with mergers, when a few large and well-consolidated manufacturers seemed to get control of the market, and giant companies from the electric power business like General Electric and Siemens started to manufacture wind turbines, this fast-increasing demand created an unexpected but welcome opportunity for new manufacturers to establish themselves. In the last few years many new companies have appeared, like Bard in Germany, Leitwind in Italy, Goldwind and Sinovel in China, to name just a few.

In China, where wind power started to grow very quickly in 2005, there are now around 20 new manufacturers of commercial wind turbines. New manufacturers and factories have also been established in Argentina, Canada, Korea and many other countries. These new companies have not started by re-inventing the wheel; know-how has been achieved through licences from engineering companies or from joint ventures.

From 2011 there has been a better supply and demand balance. The global financial crisis in 2008 led to the cancellation of some very large projects, mainly in the United States, but the market has recovered. Some manufacturers in Asia are also starting to market their wind turbines, at very competitive prices, on export markets. Wind turbines are getting larger and more cost-efficient, and the competiveness of wind power is increasing.

Offshore wind power

Denmark was the pioneer also with offshore wind power. This rather small country had made a plan for wind power development in the early 1980s, and according to this plan offshore installations should have been built in the early 1990s. A first offshore wind farm was installed at Vindeby, and a few years later a second one at Tunö Knob. Based on this experience a plan was made to install several very large offshore wind farms so that Denmark could get 50 per cent of its electric power from the wind by 2030. The reason for this was that available land for wind farms on shore was limited, and that the wind resources at sea were far better than on land. The cost of installing wind power offshore is 40–50 percent more expensive, but the increase in production will be on the same level.

During the last years several offshore wind farms have been built, Horns Rev with 80 wind turbines of 2 MW each, and Nysted which is the same size, and both have been expanded with a second phase of development. The largest offshore wind power plant so far has been installed in 2013 at Anholt, and consists of 111 wind turbines of 3.6 MW, with a total capacity of 400 MW, which delivers 4 per cent of Denmark's electric power.

Several offshore installations have also been built in Sweden, the Netherlands, United Kingdom and Germany. In 2014/15 the UK had by far most offshore wind power plants on line: 4,494 MW, which was more than half of all offshore wind power (8,771 MW) at that time (GWEC, 2015). Many countries in Europe have started a large-scale utilization of wind power offshore, with installations that can produce as much power as a nuclear power plant. The most recent 5–6 MW wind turbines have been developed mainly for the offshore market.

Re-powering

Although wind power is a new branch, many of the wind turbines installed in the 1980s and early 1990s are getting old and need to be replaced. This process is called *re-powering*. By replacing groups of small wind turbines at good sites by fewer and larger ones, production can increase significantly. On the Swedish island of Gotland six 3 MW turbines have replaced 17 old wind turbines (150 and 500 kW wind turbines). This reduced the number of turbines by two-thirds, but these six new ones produce four times more energy. Similar re-powering schemes have been conducted also on the island of Fehmarn in Germany and several other sites, and will increase in coming years.

In the 1980s and 1990s wind power was considered to be an alternative energy source which only could play a marginal role for the energy supply. Today wind power has gained market acceptance, and is a competitive energy source successfully competing on the world market.

Prospects for the future look very bright. Besides Denmark, Germany and Spain, also China, USA and India have developed a domestic wind power industry, and are developing large amounts of wind power. During coming years installations offshore, where wind resources are better and where very large wind power plants can be built, are expected to grow even further. Today several wind power plants with hundreds of multi-MW wind turbines and installed capacities of 1,000 MW or more, which can produce as much electric power as nuclear reactors, are developed, on land as well as offshore.

References

Earnest, J., & Wizelius, T. (2011) *Wind Power Plants and Project Development*. New Delhi, India: PHI Learning.

EWEA (2014) *The European Offshore Wind Power Industry: Key Trends and Statistics, 2013*. Brussels: EWEA. Accessed 2 December 2014 at http://www.ewea.org/fileadmin/files/library/publications/statistics/European_offshore_statistics_2013.pdf

EWEA (2015) 'Wind in Power, 2014 European Statistics'. Accessed 7 March 2015 at www.ewea.org/statistics/european/

Gipe, P. (1995) *Wind Energy Comes of Age*. Chichester: John Wiley.

GWEC (2015) 'Global Wind Statistics'. Accessed 7 March 2015 at www.gwec.net/wp-content/uploads/2015/02/GWEC_GlobalWindStats2014_FINAL_10.2.2015.pdf

Hills, R. L. (1996) *Power from Wind*. Cambridge: Cambridge University Press.

WWEA (2013) *Wind Energy International 2013/14*. Bonn: World Wind Energy Association.

WWEA (2014) *World Wind Energy Report 2013*. Bonn: World Wind Energy Association.

2 Wind and power

Winds are driven by the sun. Solar radiation heats the earth's surface. Since the earth is round, the angle of the solar radiation that hits the earth varies. The earth rotates around its axis, so that the radiation differs during day and night. This creates differences in temperature on different parts of the earth, and this in turn creates differences in atmospheric pressure. These tend to be equalized by movement of air from regions with high pressure to regions with low pressure. And there are the winds.

The earth is divided by longitudes that circle the earth and pass through the north and south poles, and latitudes, circles parallel to the equator. During a year the earth circles round the sun once. The sun moves (seen from a position on earth) from the southern to the northern tropical circle and back again. If a longitude is followed the solar radiation will be perpendicular to the equator (at the spring and autumn solstice). On the northern hemisphere, the angle between the sun and the surface of the earth declines the further north you move, so that the same amount of radiation will be spread over an ever larger area. Therefore the sun will heat the surface much more at the equator than at the Arctic Circle.

The rotation of the earth has a similar effect. The angle of the sun increases from sunrise until noon, and thus also the solar radiation per unit of area, and will diminish in the same way from noon to sunset. During nights the earth is not heated at all; in the night time some of the heat that has been stored in the ground and the sea radiates back into the atmosphere and returns out to space.

The earth's axis is also inclined relative to the plane where the earth moves around the sun. This gives us different seasons. At spring and autumn solstice the sun is perpendicular to the equator; day and night have equal duration. From spring solstice to midsummer the sun moves (from a position on earth) north to the Tropic of Cancer, and returns to a position over the equator at the autumn solstice and moves on south to the Tropic of Capricorn at the winter solstice, and then the sun again moves towards the equator (see Figure 2.1).

Several other factors also influence changes in temperature. Oceans cover a large part of the earth and water has quite different properties for storing heat than solid ground surface. Solar radiation can penetrate at least

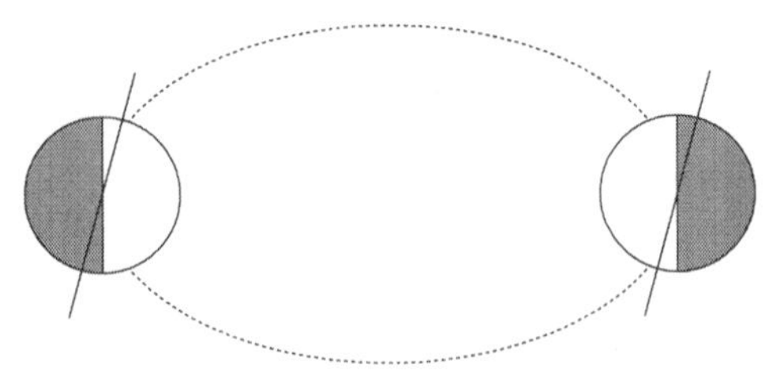

Figure 2.1 Changing seasons. The figure shows the position of the earth in relation to the sun in July (aphelion) to the left, and in January (perihelion) to the right. Since the axis of the earth is inclined, the northern hemisphere will get more sunshine in the summer than in the winter (source: author after Bogren et al., 1999)

ten metres down into water. Water stores heat better than soil; it is heated more slowly but cools slowly as well. On land the sun only warms a few centimetres of the soil, and heat is released quickly during the night. Oceans and lakes therefore have a balancing influence on temperature. Ocean currents as well as winds transport heat between different latitudes and even out the differences in temperature. The climate close to oceans is called a *maritime* climate and is quite different from the climate in the interior of large continents with a *continental* climate.

Finally there are the clouds. In the daytime the clouds shelter the earth's surface from direct sunlight and reduce the heating of the ground. In night-time the clouds reflect heat radiation back towards earth, keeping it warm. Night temperatures are warmer in cloudy weather than under clear skies.

Air mass movements

These ever-changing temperatures create more or less regular patterns for the movement of air in the atmosphere. Winds start when air begins to move from areas with high atmospheric pressures to areas with lower pressure. The wind always moves from high pressure (H) to low pressure (L).

There are winds that vary with season, like the trade winds and monsoons. In the northern middle latitudes, the wind climate is characterized by moving low-pressure areas, *cyclones* that are created in the Atlantic and move across the British Isles and Scandinavia. These low-pressure areas are created where the hot tropical and the cold polar air masses meet.

In India and other regions of Asia the wind climate is characterized by the monsoons. The summer monsoon, from southwest, starts in May/June. By that time of year the huge landmasses have been heated and are hotter than the sea, which drives this monsoon. Warm, humid air from the Indian Ocean and the southwestern Pacific moves northward and northwestward into Asia bringing heavy rain and winds. In winter a strong high pressure dominates Asia and there is an outward flow of air, reversing that of the summer monsoon. It brings dry, clear weather for a several months. The winter monsoon starts in October and goes on until March.

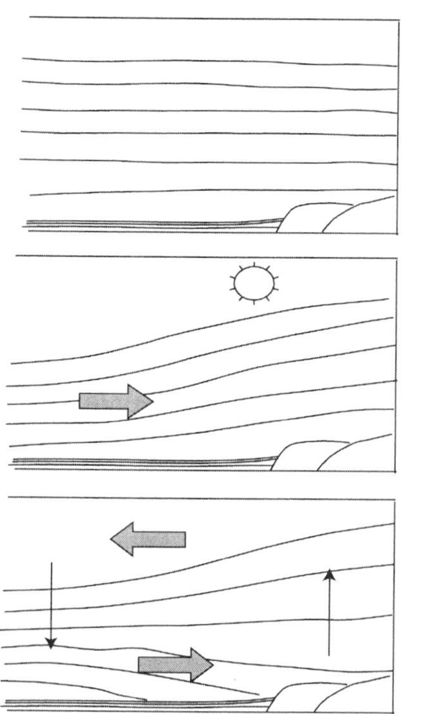

Figure 2.2 Sea breeze. During a sunny day the ground heats faster than the sea. The air temperature inside of the coast increases, air volume grows and air that is lighter starts to rise in the atmosphere, decreasing the atmospheric pressure at ground level. The cool air over the sea starts to move towards land to even out this difference in air pressure (the pressure gradient). At a certain height above ground level, atmospheric pressure is higher above land than over the sea. At that height the air starts to move the other way, towards the sea. This creates a local circulation. In the evening the ground temperature decreases faster than the sea temperature. The same mechanism creates a circulation in the other direction, the land breeze. Temperature differences between land and sea create these local winds. (source: author after Bogren et al., 1999)

In coastal areas there are local winds. The sea breeze and the land breeze give a good illustration of how winds are created (see Figure 2.2). Weather systems where air circulates between high and low pressure areas occur in the global weather system (trade winds), regional systems (monsoons, cyclones) and in local systems (sea and land breezes).

Cyclones

On a flat earth which didn't rotate and which was perpendicular to the sun, pressure differences would be equalized very quickly. Since the earth is a round rotating globe, the winds don't move along straight lines from high- to low-pressure areas. The earth's rotation creates forces that make the winds move toward the low-pressure areas in a spiral, a so-called *cyclone*.

The difference in air pressure between two areas creates the *pressure gradient force*. On meteorological maps the atmospheric pressures (with a unit hPa –hektoPascal) are illustrated by *isobars*, lines that connect points with the same air pressure. Around a low-pressure zone the isobars form irregular circles. Where the isobars are tight, the pressure gradient is strong and wind speeds higher.

Geostrophic wind

Gravity pulls air towards the earth's surface. Close to the earth, the wind is influenced by *friction* against the surface. At a specific height, this influence is negligible. This wind undisturbed by friction is called *geostrophic wind*. The distance from the ground to geostrophic wind varies and depends on weather conditions and surface roughness. High friction increases the height of geostrophic wind.

The force that makes the trade winds blow towards the equator from northeast and southeast instead of in a perpendicular direction is called the *Coriolis force*. This force diverts the movement of the air to the right on the northern hemisphere. The Coriolis force is always perpendicular to the direction of movement. It is proportional to the wind speed and varies with the latitude; at the equator it is zero and then it increases towards the poles. Another better-known force acts in the same direction (but is not as strong); the *centrifugal force*.

A wind that blows from the northern hemisphere perpendicular to the equator is diverted to the right, so that the wind will come from northeast. Winds that blow north from the equator, will also be diverted to the right, and will come from southwest. The monsoon winds follow these directions.

If there is a low-pressure zone in the north and a high-pressure zone in the south, with parallel isobars from west to east, the pressure gradient force will set the air in motion. The wind will blow from south to north and a so-called air package will accelerate northward. The pressure gradient pulls the air perpendicular to the isobars. When wind speed increases, the Coriolis force will divert the moving air towards right (east). The higher the wind speed, the more it will be diverted. Finally a balance will be attained where the pressure gradient force and the Coriolis force will be equal. The wind where these forces are in balance is called the *geostrophic wind*. This definition corresponds well with the wind undisturbed by friction.

Wind profile

At lower heights the friction against the earth's surface influences the wind. The friction force acts in the direction opposite to the air movement. The friction retards the wind and creates an imbalance between the pressure gradient force and the Coriolis force; the wind moves across the isobars (takes a shortcut to the low pressure centre). The friction will be stronger closer to the ground surface. The wind speed will decrease the closer to the surface you get. At the same time wind direction will change, more across the isobars closer to the surface. This change of wind speed and wind direction is called *wind shear.*

Wind turbines have become higher and higher, but they will always stay within the so-called *friction layer* of the atmosphere. For wind power, winds up to 200 metres above ground (or sea) are of interest. Within these heights the wind is always influenced by local conditions: the terrain at the actual site and in an area with approximately 20 km radius around the site. The character of the earth's surface will influence the strength of the friction force. This force is of course lower over an open field than over a hilly forest area.

At the earth's surface, the wind speed is always zero. This does not mean that the air doesn't move, but that the sum of its movements in different directions adds up to zero.

Wind speed increases with height. How fast this increase will be depends on the friction against the earth's surface. The wind will not be retarded very much on an open plain with low friction, and the increase with height will not be very large. The relation between wind speed and height is called the *wind profile* or *wind gradient* (see Figures 2.3 and 2.4).

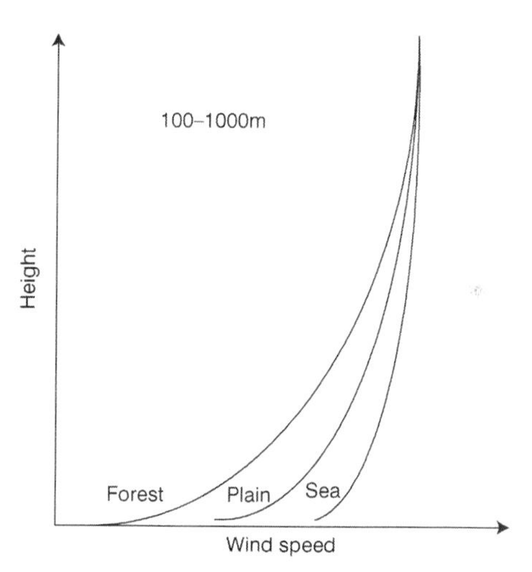

Figure 2.3 Wind profile. The wind profile, or wind gradient, describes the relation between wind speed and height above ground level (agl). The form of this profile depends on the friction of the earth surface. If the surface is very smooth (water) the friction against the ground has a moderate effect, and the wind profile will be almost vertical (right-hand line on the graph). If the surface is rough the wind near the surface will lose speed and the wind profile will bend. Over a plain or areas with agricultural fields, the friction will be stronger, and the graph will be more curved (middle line). Over a forest the friction will be very strong, and the profile will be more curved that over a plain (left-hand line). However, the wind speed over a rough surface does not catch up with the wind speed over a smooth surface until a height of 100–1,000 metres

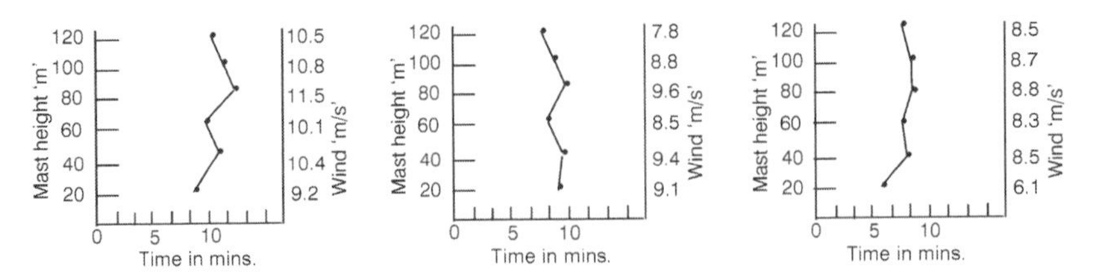

Figure 2.4 Real-time wind profile. The graph from a measuring mast with anemometers at different heights, in real time, does not look so even and smooth as the lines in Figure 2.3. These diagrams show real-time measurements at 1-minute intervals. It shows that wind speed can be higher at for example 40 m than at 60 m and that the form of the profile will change continuously. The wind profile describes a long-term average of the relation between wind speed and height. At a specific site the wind profile will also look different from different wind directions, since it is the character of the terrain in the sectors that the wind comes from that forms the wind profile

When an area is covered by forest, the wind profiles do not start at the ground level, but at three-quarters of the height of the trees. This distance is called the displacement height.

Turbulence

Since air is an invisible gas, it is hard to observe how air moves. It is easier to observe what happens when a stream of water meets an obstacle; around a stone whirls and eddies are formed. The air reacts in a similar way. When the wind hits an obstacle whirls or waves are formed. When the air moves parallel to the ground, it is called *laminar* wind. When it moves in different directions around the prevailing wind direction, in waves and eddies, the wind has become *turbulent*.

The waves in the air can have a wavelength of several hundred metres. Whirls of air can also be very big, but will successively be broken down into smaller eddies and finally to movement on the molecular level; they turn into heat. When wind is measured, these waves and whirls appear as short variations of wind speed: turbulence.

Temperature differences in the air can also create turbulence and reduce the wind speed. If the air close to ground is warmer than on higher levels, and the temperature decreases rapidly with height, warm air will rise upwards. The horizontal wind will then meet air that moves in a vertical direction, and this creates turbulence.

When the wind passes through the rotor of a turbine, a very strong turbulence is created. This whirling wind on the lee side of the rotor is called *wind wake* and influences the wind speed up to a distance of 10 rotor diameters or more behind the turbine.

The turbulence is measured as *turbulence intensity* (I). Since turbulence increases with wind speed, the speed has to be annotated as well: I_{15} for 15 m/s.

The turbulence intensity is the quotient of the standard deviation and the 10-minute average wind speed:

$$I_u = \frac{\sigma}{u_{average}}$$

The standard deviation is the root mean square$[u_{average} - u_n]$.

Local impacts on the wind

Some hundred metres up in the sky, the wind moves uninterrupted, driven by the forces described in the previous section. Closer to ground, at the height where the wind turbine rotors are turnin, the local terrain affects the wind. The orography, the height contours, compresses the air or creates turbulence. The ground cover can be smooth or rough, and obstacles like buildings, trees and other structures also interrupt the movements of the air.

Internal boundary layer

When the wind blows from the sea over land, the friction against the surface will change at the shoreline. The air that moves close to the surface will start to rotate. This turbulent air will move upwards when the wind moves further over land. An *internal boundary layer* has been created. The wind speed below this boundary layer (the boundary between the laminar wind from the sea and the more turbulent wind over land) will be retarded, while the height of the boundary layer will stabilize and the wind gradient will change into a new form. If the wind passes from land to the sea, the air close to the water surface starts to accelerate instead.

Each time the character of the landscape and thus the surface friction changes, a new internal boundary layer is formed that will change the wind speed to a level that corresponds to the friction of the underlying terrain. It always takes some time and distance before this change of friction has been transferred to higher layers. The wind energy content and changes of the surface friction overlap (see Figure 2.5).

Roughness of terrain

The terrain is classified into different *roughness classes*. Five different classes are used, from 0 to 4. Open water is 0, open plain 1 and so on to roughness class 4 for large cities and high forest (see Table 2.1).

Sometimes the expression *roughness length* is used. This length has nothing to do with the length of the grass or the height of buildings, it is a mathematical factor used in the algorithms for calculations of how the terrain influences the wind speed.

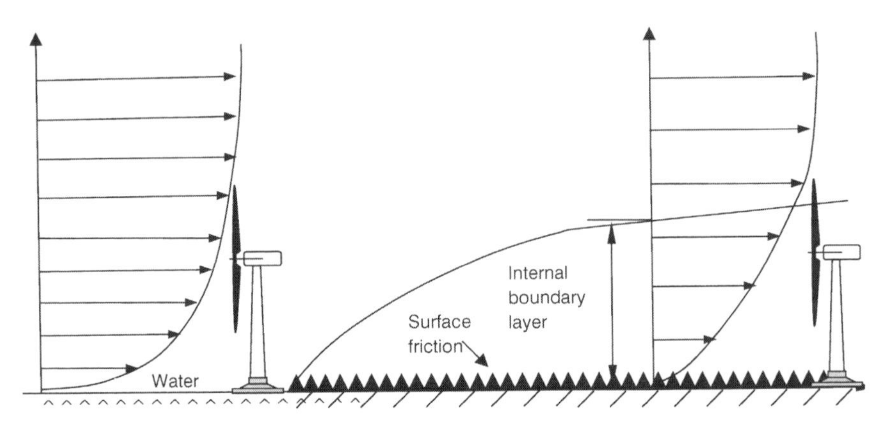

Figure 2.5 Internal boundary layer. The wind is affected by the friction against the surface. When the wind blows from the sea (left) to land (right), turbulence will increase and the wind speed close to the ground will decrease. Between the turbulent wind and the laminar wind from the sea at higher levels an internal boundary layer is formed. The height of the boundary layer will stabilize after a while, and remain until the surface roughness changes again. The wind profile will change; over land (right) it is more curved than over sea (left). The three arrows close to the surface, which represent wind speed at these three heights, are shorter over land than over sea. Compare this also to the wind profiles in Figure 2.3.

Table 2.1 Roughness classes

Roughness class	*Character*	*Terrain*	*Obstacles*	*Farms*	*Buildings*	*Forest*
0	Sea, lakes	Open water	–	–	–	–
1	Open landscape, with sparse vegetation and buildings	Plain to smooth hills	Only low vegetation	0–3 per km^2	–	–
2	Countryside with a mix of open areas, vegetation and buildings	Plain to hilly	Small woods, tree-lined roads are common	Up to 10 per km^2	Some villages and small towns	–
3	Small towns or countryside with many farms, woods and obstacles	Plain to hilly	Many woods, vegetation and tree-lined roads	Many farms more than10 per km^2	Many villages, small towns or suburbs	Dense forest (closed canopy)
4	Large cities or forests	Plain to hilly	–	–	Large cities	Forest (open canopy)

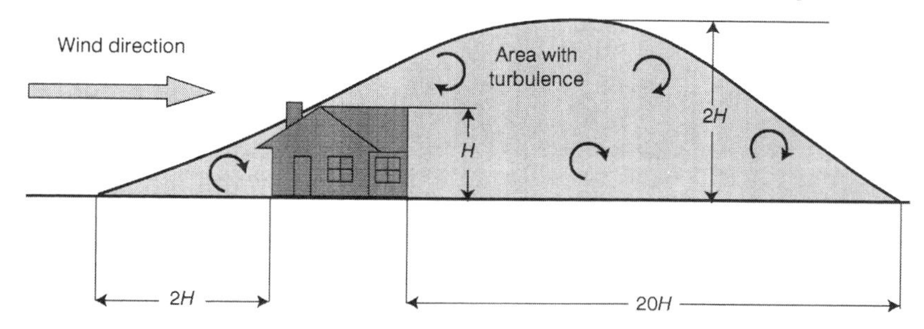

Figure 2.6 Turbulence from obstacles. Close to an obstacle the turbulence will increase and the wind speed will decrease. The turbulence is spread further on the leeside of an obstacle, but also on the side from where the wind comes turbulence will appear, since the obstacle interferes with the airflow. The areas with turbulence will of course vary with the wind direction (illustration: S. Piva after Gipe).

Obstacles

Obstacles, like buildings, trees and other high structures, affect the wind. They will interrupt the laminar flow of air and create whirls and eddies. This will reduce the wind speed. How much impact an obstacle has depends on not only the height and width of the obstacle, but also on the porosity: how much the wind that can pass through the obstacle. In a garden, a hedge or a fence that allows some air to pass through is a much more efficient windbreak than a brick wall or a dense fence. When the wind hits a compact wall strong turbulence is created behind it. If some of the wind can pass through the fence, there will be less turbulence. An obstacle, be it a building, tree-lined road or a wood, will influence the wind in front, behind and above the obstacle. According to a simple rule of thumb, an obstacle creates turbulence to double the height of the obstacle, starting at a distance of twice the height in front of it and continuing 20 times the height behind the obstacle (see Figure 2.6).

Obstacles have to be quite close to a wind turbine to have a negative impact on its production. In harbours and industrial areas, where there are silos, oil storage tanks and other large structures, this has to be taken into account when a site for wind power plants is chosen. A more exact estimation of how the wind speed is influenced at different distances and heights from an obstacle can be made with a diagram from the European Wind Atlas (see Figure 2.7).

Slopes and hills

Slopes and hills can also have an impact on the wind speed. The form of the terrain can create a so-called hill effect, so that the wind speed will increase up to a certain height above ground level. When the wind passes over an even hill, the airflow will be compressed so that the wind speed over the top of the hill increases.

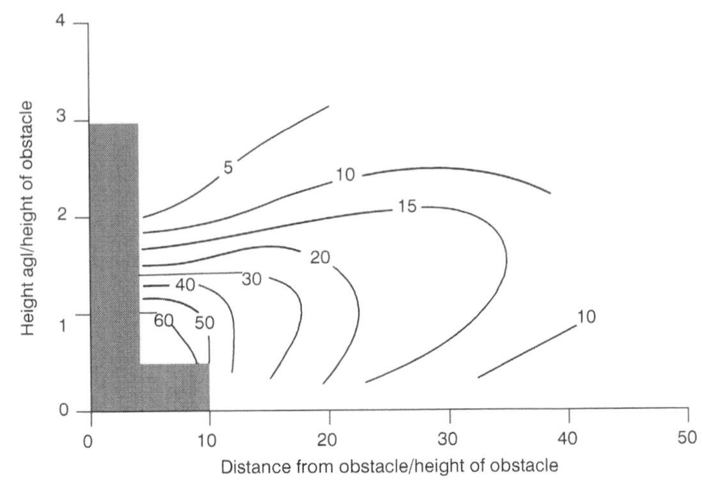

Figure 2.7 Diagram for estimating impact of obstacles. This diagram gives the relation between an obstacle and the reduction of wind speed at different heights and distances from an obstacle. The vertical axis shows the heights above ground level divided by the height of the obstacle. The horizontal axis gives the distance from the obstacle divided by the height of the obstacle. (Note: the rectangular area at the *y*-axis is not the obstacle, but the area too close to the obstacle to make measurements/ calculations.) (Source: Troen and Petersen, 1989)

This effect depends on local circumstances. The increase of wind speed is largest on moderate heights, 20–40 metres. A smooth but not too steep hillside will make the wind speed increase, up to a certain height above the hill. Behind the hilltop, on the down slope, the wind speed will decrease again. However, if the hillside is too steep, the airflow becomes turbulent and the effect will be the opposite, a decrease in wind speed (see Figure 2.8).

Complex terrain

Terrain with mountains, deep valleys and steep slopes is called *complex terrain*. When the airflow moves over this kind of terrain, special phenomena occur. This makes it hard to predict the wind speed without measurements at the site. Mountain valleys are interesting examples. Along the sides you will get local winds that are created by the temperature differences at the top of the hillside and the bottom of the valley, and that will vary in day and night. In the middle of the valley a so-called tunnel effect can occur; strong winds that follow the direction of the valley, with a maximum wind speed at a relatively low height above ground.

A wind power project developer is first of all interested to find out how much energy the wind will contain during the technical lifetime of a wind turbine, 20–25 years. To get good production and economic returns, the

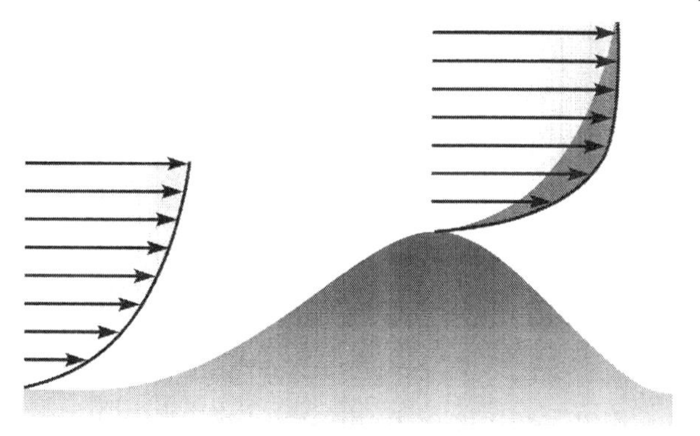

Figure 2.8 Increase of wind speed on a hill. When wind is passing over a smooth hill, wind speed will increase to the hilltop. To get this effect the inclination of the slope of the hill should be less than 40 degrees, and if the hillside is uneven and rough, or covered with trees, the wind flow can be disturbed at inclinations of even 5 degrees. On the leeside of a hill the wind speed will decrease. The wind can accelerate around the sides of hill in a similar way. The marked area of the wind profile (to the right) shows the increase compared to the wind profile in front of the hill (to the left). (illustration: S. Piva after Troen and Petersen, 1989)

wind power plants have to be sited in areas where the wind contains much energy. To find the right site in the terrain, it is also necessary to know something about the wind directions. It is crucial to get a realistic idea of the *wind climate* in the area where wind power plants will be installed (see Box 2.1).

For a wind power developer good information about the *local wind climate* is most important.

Power in the wind

Wind is air in motion. Since air has mass, the wind contains kinetic energy (the weight of air is a little more than one kilogram per cubic metre). This power can be turned into electric power, heat or mechanical work by wind turbines. The wind can be very strong and powerful. It can be strong enough to break off trees and rip off roofs from buildings. Storms and hurricanes can create natural disasters. The power of the wind can be so strong since it is proportional to the *cube* of the wind speed. When the wind speed doubles, the power increases eight times.

Even a small increase of the wind speed, for example from 6 to 7 m/s, gives a large increase in power; in this case, it will be increased by ~60 per cent (see Box 2.2 and Figure 2.9).

Therefore wind turbines should be installed at sites with the best possible wind resources.

Box 2.1 Wind climate

Weather is the instantaneous state of the atmosphere: temperature, atmospheric pressure, wind speed, humidity, cloudiness, rain, sunshine and visibility. It is the totality of atmospheric conditions at any particular place and time.

Climate is a wider concept; it is the sum of the weather at a place over the years. Since the average conditions of the weather change from year to year, climate can only be defined in terms of some period of time, a run of years, a particular decade or some decades.

The *wind climate* is the long-term pattern of the wind in a specific site, region or country.

The climate can be studied and analysed on different levels.

- the *macro climate* is the large-scale climate patterns on the earth, continents or parts of continents;
- the *meso climate* is the climate in a country or a region; and
- the *local climate* is the climate within a limited area such as a coastal zone, a wood or a field.

Power density and energy content

Wind turbines transform the power in the wind into usable energy. They should be sited where the wind is powerful and contains a lot of energy.

The *power* of the wind at a specific height at a site is usually specified as *power density* (W/m^2) and the energy content as kWh per square metre and year (kWh/m^2 per annum); the energy in the winds that pass through a vertical area of one square metre during one year. Power density and energy content are in fact two different ways to express the same thing. If the power density is known, the energy content is easy to calculate. The power density (W/m^2) is multiplied with the number of hours in a year (365 days × 24 hours = 8,760 hours) and divided by 1,000 to get the kWh/m^2 per annum (see Figure 2.9).

It is important to distinguish the concepts *power* and *energy* (see Box 2.3). Some wind resource maps just show the average wind speed. Modern wind resource maps give information about the power density or energy content in different areas by isolines or by colours, which gives much better information than maps just showing average wind speeds.

Frequency distribution of wind speeds

At a site where the wind speed always is exactly 6 m/s, the power density of the wind will be given by:

Box 2.2 The power of wind

The power of the wind is calculated in the following way:

$$P_{kin} = \frac{1}{2}\dot{m} \cdot v^2$$

where

P_{kin}= kinetic power W (J/s)

$\dot{m}$ (mass flow) = $\rho \cdot A \cdot v$

ρ = air density (kg/m³)

A = area (m²)

v = wind speed (m/s)

Air is not a solid mass but a fluid, so $\dot{m}$ can be substituted by $\rho \cdot A \cdot v$, then the equation above becomes

$$P_{kin} = \frac{1}{2}(\rho \cdot A \cdot v)v^2$$

$$\text{i.e. } P_{kin} = \frac{1}{2}\rho \cdot A \cdot v^3$$

The density of air varies with the height above sea level and temperature. The standard values used are usually density at sea level (1 bar) and temperature 9° is 1.25 kg/m³. Then the power of the wind per m² is:

$$P_{kin} = \frac{1}{2}1.25 \cdot v^3 = 0.625v^3$$

This indicates that the power is proportional to the *cube* of the wind speed.

If $v = 4\text{m/s}, \ P = 0.625 \times 4^3 = 0.625 \times 64 = 40\text{W}$

If $v = 8\text{m/s}, \ P = 0.625 \times 8^3 = 0.625 \times 512 = 320\text{W}$

$8^3 = (2 \times 4)^3 = 2^3 \times 4^3 = 8 \times 4^3; \quad 320 / 40 = 8$

When the wind speed doubles, the power increases by a factor 8.

Power density = 0.625 × 6³ = 135 W/m²

and the

Energy content = 0.625 × 63 × 8,760 = 1,182 kWh/m2 per annum

In the real world the speed and direction of the wind change continuously. Some days are calm, other days storms are windy. The wind will change during day and night, at different seasons and from year to year. To find and estimate the power density or energy content at a site, the *average* power density has to be calculated.

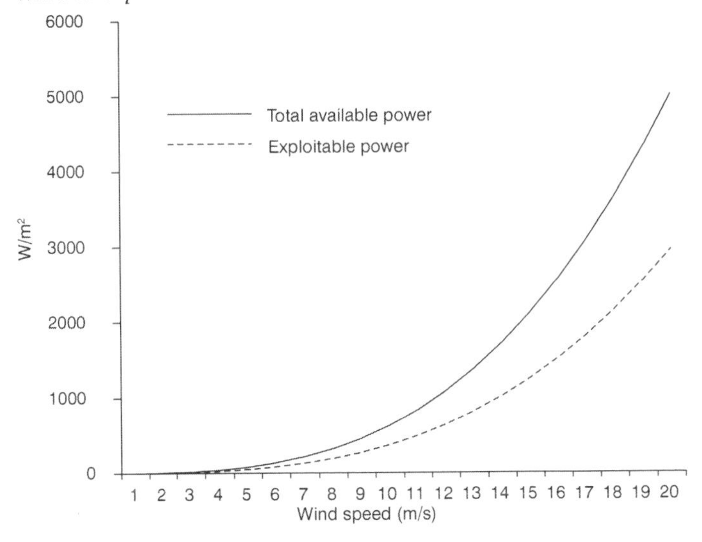

Figure 2.9 The power of the wind. The power of the wind is proportional to the cube of the wind speed. The power is expressed as the power density (W/m^2). A wind turbine can theoretically utilize 59 per cent of the power in the wind (according to Betz' law), shown by the curve to the right

> *Box 2.3* Power and energy
> *Power* is energy per time unit and is expressed in Watts (or kW, MW, GW). Power is often notified by the letter *P*. 1 W = 1 J/s (joule per second).
>
> *Energy* is power multiplied by the time the power is used. A wind turbine which gives 1,000 kW power during one hour has produced 1,000 kiloWatt hours (kWh). If the wind turbine gives 300 kW power on average during a year, it produces 300 kW x 8,760 hours = 2,628,000 kWh per annum.
>
> Power is energy per unit of time. Energy is power multiplied by time.

If the wind speed (v) is measured at a site during one year, and the wind speeds are sampled on regular intervals, it is easy to calculate the *mean wind speed* for that site, by adding all the measured values for the wind speed (Σv) and then divide the sum by the number (n) of observations:

$$v_{\text{mean}} = \frac{\sum v_n}{n}$$

At a site where the mean wind speed is 6 m/s, it could be assumed that the power density would be 135 W/m^2 and the energy content 1,182 kWh/m^2, considering the calculation made above. Unfortunately this assumption is false, since the power of the wind is proportional to the *cube* of the wind speed.

The cube of the sum of the wind speeds $(v_1 + v_2 + v_3 + \ldots + v_n)^3$ is not the same as the sum of the cubes of the wind speeds $(v_1^3 + v_2^3 + v_3^3 + \ldots + v_n^3)$.

If the wind speed at a site is 4 m/s for half of the year and 8 m/s for the rest of the year, the mean wind speed (v_{mean}) will be:

$$v_{mean} = 4/2 + 8/2 = 6 \text{ m/s}$$

The power density will be: $\frac{1}{2}(4^3 + 8^3) = \frac{1}{2}(64 + 512) = 288$ W/m^2 and the energy content: 0.625 × 288 × 8,760 = 1,576 kWh/m^2 in a year.

To calculate the power density and energy content of the wind at a site it is not sufficient to know only the mean wind speed. It is necessary to know all the different wind speeds that occur and their duration; the *frequency distribution* of the wind speeds has to be found. The power density of the wind at two different sites with exactly the same mean wind speed can differ considerably.

Data on wind speeds are sorted into a bin-diagram, with wind speed on the x-axis and the duration (in hours or percent) on the y-axis (see Figure 2.10). To calculate the energy content of the wind during one year, the cubes of the wind speeds are multiplied by the frequency, the products are added up and the sum is entered into the formula above.

In the example above, 1,182 kWh/m^2 per year has to be multiplied by 1.33 to give the correct answer. This factor is called the *cube factor* or *energy*

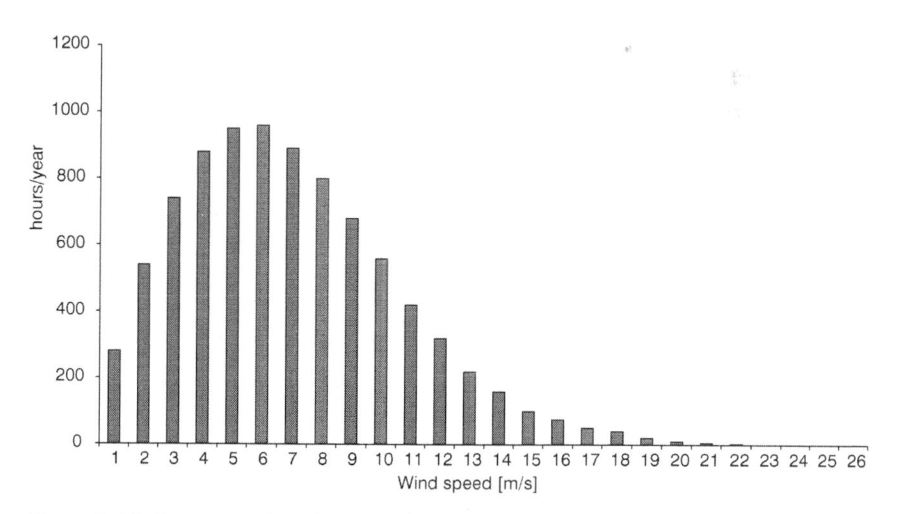

Figure 2.10 Frequency distribution of wind speed. Frequency distributions of wind speeds can look like this. The most common wind speeds are 5–6 m/s. During ~950 hours a year, 11 per cent of the time, the wind speed is 6 m/s

pattern factor (EPF), and can be added to the following formula to calculate the energy content of the wind during a year (E_{year}):

$$E_{year} = 0.625 \times v^3 \times 8{,}760 \times \text{EPF}$$

The value of the cube factor depends on the frequency distribution of the wind. If the mean wind speed is known but the frequency distribution is not known, a cube factor of 1.9 will in most cases give a good idea of the energy content at a site. This however applies only to places in mid latitudes: the United States and most parts of Europe. In areas with trade winds or seasonal winds, or dominant local winds, other values for the cube factor should be used. In Puerto Rico, with trade winds, the cube factor is 1.4; in the San Gorgonio pass in California, with local winds it is 2.4; and in southern India the cube factor is 1.8.

The Weibull distribution

There is a statistical probability distribution called the *Weibull distribution*, named after the Swedish scientist Waloddi Weibull. This distribution was originally created to describe fatigue loads within mechanical engineering. It has been shown that the Weibull distribution also incidentally fits quite well to the frequency distribution of the wind, and it is widely used in this context. The frequency distribution at a specific site contains a large amount of data. These data will be transformed into the Weibull parameters (see Figure 2.11).

The Weibull distribution is a statistical model defined by two parameters, which can be altered to change the scale and the shape of the curve. The

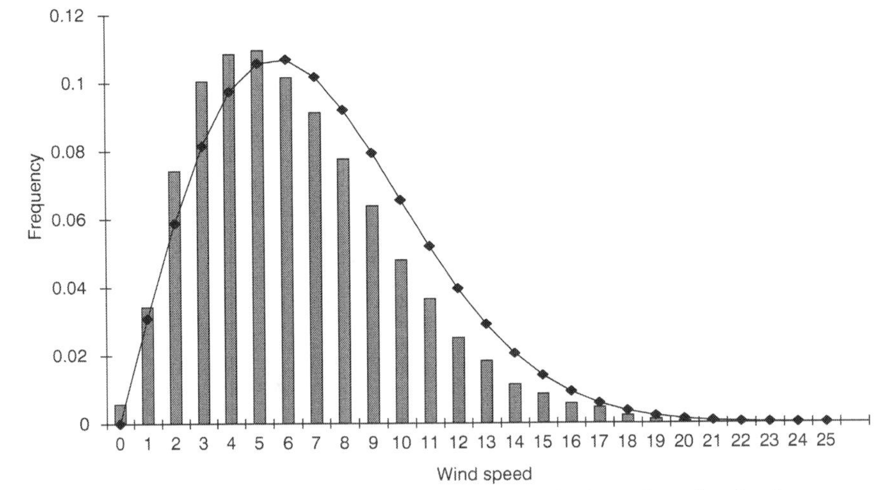

Figure 2.11 Weibull fit of frequency distribution. The bins describe the frequency distribution and the curve is the best Weibull fit to this distribution

area under the curve is always the same. It describes the *probability* of the occurrence of different wind speeds (x-axis)(see Figures 2.12 and 2.13).

A 1 MW turbine with a nominal wind speed of 14 m/s, installed at a site with an average wind speed of 6.2 m/s, will produce around 10 per cent

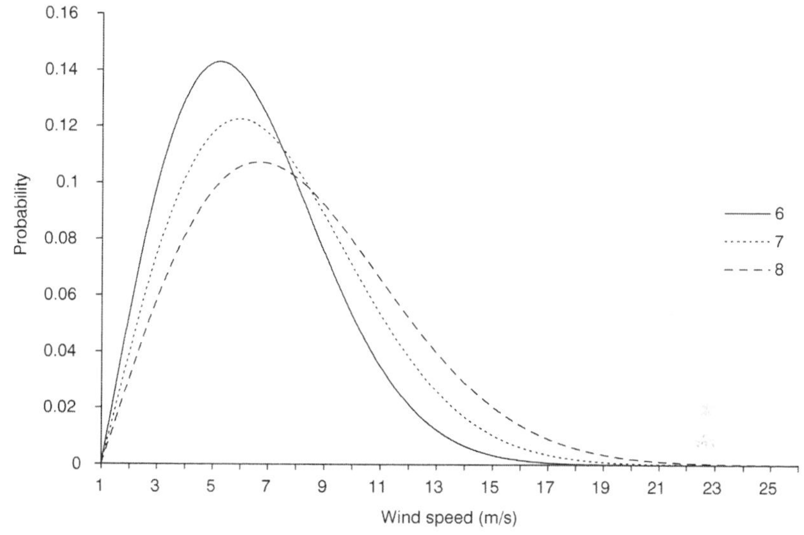

Figure 2.12 Weibull parameters. The Weibull distribution is defined by two parameters scale A and shape k. In this diagram the shape parameter is k=2 (the so-called Rayleigh distribution). The average wind speeds are 5.3 (solid), 6.2 (dotted) and 7.3 (dashed), and scale parameters 6, 7 and 8 respectively

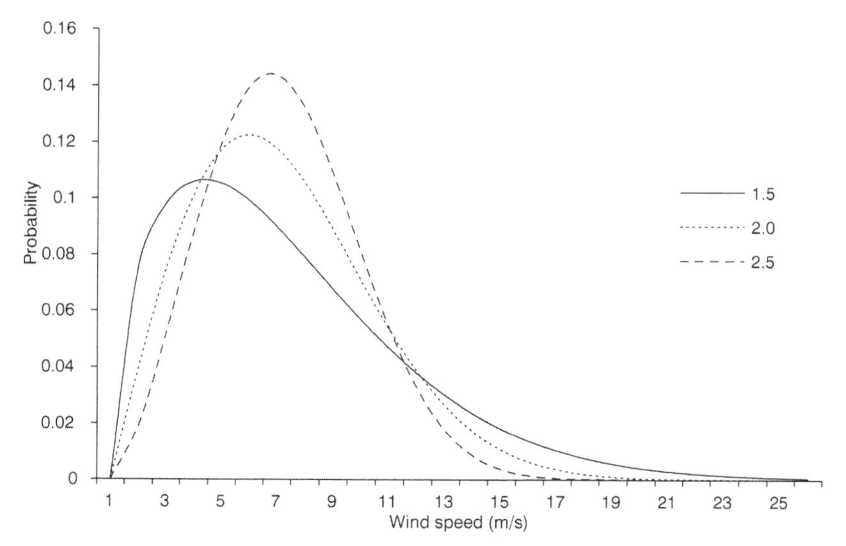

Figure 2.13 Weibull parameter k. In this diagram the scale parameter A is 7, and the average wind speed 6.2 m/s. The shape parameters are 1.5 (solid), 2.0 (dotted) and 2.5 (dashed)

more when the shape parameter $k = 1.5$ than for $k = 2.5$, although the average wind speed is the same.

The power density of the wind (energy content) at two different sites with exactly the same mean wind speed can differ considerably. This is due to differences in the frequency distribution of the wind.

The relation between wind speed and height

As a general rule wind speed will increase with height. How large this increase will be depends on the roughness of the terrain. In areas with high roughness, the wind speed will increase more with height than over a smooth terrain. But the wind speed at a specific height, for example 50 m agl, will always be higher in an area with low roughness, if all other factors are equal.

For wind turbines it is the wind speed at hub height that is of interest. This height varies for different models and manufacturers. Available wind data usually represent a different height than the hub height. It is however not very difficult to recalculate these data for other heights (see Box 2.4).

Long-term wind climate

Most wind turbines have a technical life of 20–25 years. From the data collected by wind measurements the wind speed and the frequency distribution during the *coming* 20 years has to be estimated. This prognosis has to be based on solid assumptions.

If the wind is measured very accurately for 12 months, what do we actually know? The only thing we know for sure is the wind's characteristics during this specific period. What conclusions can be drawn from these facts about the wind's power density in coming years?

The wind speed, frequency distribution and averages vary significantly in different years. Also long-term averages for five- and ten-year periods can vary a lot. How the power density at a site will vary in the long term is important to know if the power of wind is going to be utilized. The longer the period is compared, the less is the variation. This is reasonable from a statistical point of view. However, today when climate change no longer is just a threat but a fact, the uncertainty of prognoses for future winds has increased. The power in the wind, energy content or power density can vary by as much as 30 per cent in different ten-year periods (see Figure 2.14).

To get good background data for a prognosis, measured data for a much longer period than one year is necessary. It is however no sensible strategy to measure the wind for five to ten years before a decision to develop a wind power plant is taken. In most cases, the long-term average wind speed will not differ more than 10 per cent from a single year, in 90 per cent of the cases (90% confidence interval). In Europe the standard deviation of the

Box 2.4 Wind speed at different heights

If the average wind speed at a height (h_0) is known and you want to find the wind speed at hub height (h), the following relation can be used:

$$v / v_0 = (h / h_0)^\alpha$$

where,

v_0 is the known wind speed at the height h_0

v is the wind speed at the height h

The value of the exponent α depends on the roughness of the terrain:

- roughness class 0 (open water): $\alpha = 0.1$
- roughness class 1 (open plain): $\alpha = 0.15$
- roughness class 2 (countryside with farms): $\alpha = 0.2$
- roughness class 3 (villages and low forest): $\alpha = 0.3$

Example: If the average wind speed on an open plain (roughness class 1) is 6 m/s at 10 m agl, determine the average wind speed at 50 m agl.

Solution: Using the formula $v / v_0 = (h / h_0)^\alpha$

$$v_{10} = 6,\ h = 50\text{m},\ h_{10} = 10\text{m},\ \alpha = 0.15$$

then $v_{50} = 6(50 / 10)^{0.15} = 7.6$ m/s

There are different values for α in the literature. These values come from the wind atlas program WindPRO. This method is called the *power law.* There is another method as well, but this simple one usually gives the best results when $h \geq 50$ m.

With the wind speeds at two known different heights, the wind gradient exponent (also called *Hellman Exponent)* can be calculated by:

$$\alpha = \frac{\log(v / v_0)}{\log(H / h_0)}$$

It important to be aware that there is a high possibility of inaccuracy using this simple power law equation, especially in a complex terrain.

long-term wind speed is about 6 per cent. The power density, that is the available wind energy, will however differ much more, since the power is proportional to the cube of the wind speed.

Wind data from a site that has been logged for a shorter period have to be adapted to a so-called *normal wind year* that is an average for a period of five to ten years, before it can be used to calculate the power density at the site and the energy production of a wind power plant. The measured wind data have to be compared with corresponding data from the same measurement period in the same region, where long-term data are also available. Then it

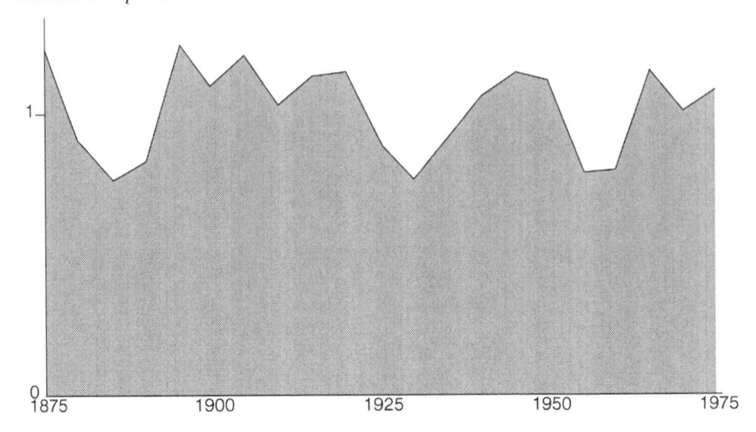

Figure 2.14 The energy content in the wind during five-year periods in Denmark. This diagram shows how the energy content of the wind has varied during five-year periods from 1875 till 1975 at Hesselö in Denmark, compared to the average for the whole 100-year period (source: Troen and Petersen, 1989; DTU Wind Energy)

can be checked how representative the data from the measurement period are, compared to the long-term data from the second measuring mast. Finally the collected wind data can be adjusted so that they will correspond to a *normal* year, the long-time average.

National meteorological institutes have collected wind data for decades from a large number of meteorological stations in different parts of their countries.

The wind atlas method

On most sites it is possible to calculate the power density and energy content of the wind without using measuring equipment at the specific site. Alternatively the wind data from an existing measuring mast for which long-term data is available can be recalculated by the wind atlas method to represent the site. In complex terrain and where available data are unreliable (in mountainous areas, large lakes and at sea), this method can't be applied and it is necessary to make on-site wind measurements.

This method of calculating the energy content at different sites was developed by scientists from the research station Risoe in Denmark in the 1980s. They made careful measurements of how the wind was influenced by different kinds of terrain, hills and obstacles. From these empirical data they developed models and algorithms to describe the influence of hills, different kinds of obstacles and orography.

These algorithms were then used in a computer program, WAsP, that can be utilized to calculate the energy content at a given site by using wind data from an existing wind measuring mast with long-term wind data (that are converted to so-called *wind atlas data*) and with information that describes obstacles, hills and the roughness of the terrain within a radius of 20 km from the site where the wind turbine will be installed.

A *wind atlas program* works in two stages. The first stage is to convert normal long-term (5–10 years) wind data (wind speed and direction) from a regular wind measurement mast to so-called *wind atlas data*. This means that the wind data from the measuring mast are normalized to a common format, so that data from different masts are comparable and can be used by the program.

Wind measurement masts often stand close to buildings and are surrounded by different types of terrain and often by hills and mountains. The program can 'delete' the influence from obstacles, orography (height contours) and terrain (roughness), so that the measured wind data are converted to what they would be if the terrain had been plain (roughness class 1) without any hills or obstacles, at 10 m above ground level.

The first set of wind atlas data consists of the frequency distribution of the wind in twelve sectors (N, NNW, WNW, etc.) 10 m agl in roughness class 1. These data are then recalculated to other heights: 25, 50, 100 and 200 metres. Together these data describe the *regional wind climate* in an area with a radius of approximately 20–100 km (the size of the area depends on local conditions) where the geostrophic winds are the same.

To calculate the energy content of the wind and how much a specific wind turbine can be expected to produce at a given site, the same procedure is followed, but the other way around. Within a reasonable distance from the measuring mast, which has been used to process the wind atlas data, the properties of the winds at 200 m agl should be the same.

By entering data about the roughness of the terrain within 20 km radius from the site, and data about hills and obstacles, and finally data about the wind turbine (hub height, rotor-sweep area, and power curve which describes how much the turbine will produce at different wind speeds), the program calculates the frequency distribution of the wind at hub height. Finally the program calculates how much the turbine can produce at that site during a normal (average) wind year (see Figure 2.15).

There are several different types of computer software for wind power applications, that are based on the *wind atlas* method. All of them are easy to work with and give reliable results, if the operator is experienced. They can be used to calculate how much a wind turbine of a specific brand/model can produce at a given site, as well as the sound propagation, shadow flicker and visual impact. They can also calculate how the impact of *wind wakes* on park efficiency (i.e. how much power the wind produces compared with if they were standing on their own) and create *wind resource maps*.

Wind wakes

If only one wind turbine is to be installed, the position of the turbine will be based on the roughness of the terrain, distance to obstacles and the height

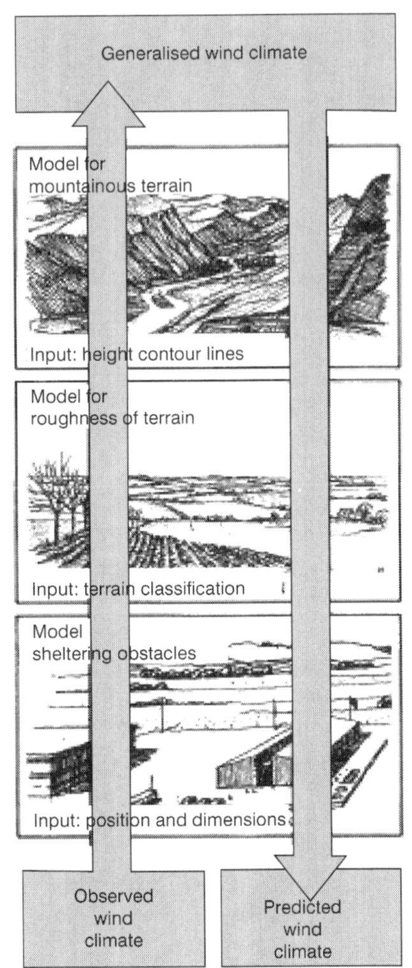

Figure 2.15 Wind atlas method. Wind data from meteorological stations are first converted to wind atlas data by clearing away the influence of obstacles, terrain roughness and heights (from bottom up, to the left). These data are then used to calculate the wind climate at specific sites, by adding the influence of the specific conditions (obstacles, roughness, heights) at that site (top down arrow to the right) (source: Troen and Petersen, 1989; DTU Wind Energy)

contours of the surrounding landscape. If more than one turbine is to be installed at a site, the turbines will also have an impact on each other. How large this impact will be depends on the distance between the turbines and the distribution of the wind directions at the site.

On the down-wind side of the rotor a wind wake is formed; the wind speed slows down and regains its undisturbed speed some ten rotor diameters behind the turbine (see Figure 2.16). This factor has to be taken into account when the layout for a group with several wind turbines is made.

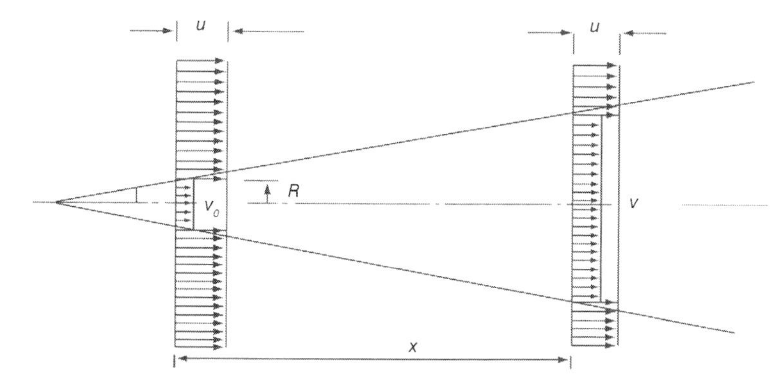

Figure 2.16 Wind wake. The wind speed (u) is retarded by the rotor (v_0). Behind the rotor the wind speed increases again (v) as the wake gets wider. R = rotor distance, X = distance from wind turbines in metres (source: Jensen, 1986; DTU Wind Energy)

The wind speed is retarded by the wind turbine rotor, and behind the rotor the wind speed increases again until it regains its initial speed. The extension of the wind wake determines how the individual turbines will be sited in relation to each other in a group of turbines. The diameter of the wind wake increases by about 7.5 metres every 100 metres downwind of the rotor, and the wind speed will increase with the distance until the wake decays completely.

The relation between the wind speed v and the distance x behind the rotor is described by the formula:

$$v = u\left[1 - \frac{2}{3}\left(\frac{R}{R + \alpha x}\right)^2\right]$$

where
v is the wind speed x metres behind the rotor
u is the undisturbed wind speed in front of the rotor
R is the radius of the rotor
α is the wake decay constant (how fast the wake widens behind the rotor).

The wake decay constant α depends on the roughness class. On land this value is usually set to 0.075 [m], offshore the value is set to 0.04 [m]. Several more advanced models for calculations of wind wakes have been developed since this was formulated in 1986 by N.O. Jensen at Risoe, but this example shows the principle of wind wakes quite well.

Wind resource maps

In many countries, meteorological institutes have converted wind data (from 5–10-year periods and longer) from a large number of measuring masts to wind atlas data, which can be used as a database in wind atlas programs. These wind atlas data can then also be used to create wind resource maps.

The European Wind Atlas was developed in the late 1980s, by joint efforts of researchers and meteorologists in the European Union. It gives an overview of the wind resources in Europe. Similar wind atlases, using the wind atlas method, have since then been developed for many countries and regions in the world. These are available at www.windatlas.dk, and many wind resource maps can be found on the internet.

In Sweden the meteorological institute SMHI produced wind resource maps for counties in the southern parts of Sweden. The energy content of the wind is shown as isolines (kWh/m^2, year) that connect points with the same energy content. Isolines for wind energy are named *isovents*.

Since the wind speed increases with height above ground level, the *height* for the wind resource map always has to be specified. The standard height for wind measurements is 10 m agl. For wind power, the relevant height for calculations is the hub height, which depends on the size of the turbine. SMHI's maps are made for two heights, 50 m and 80 m agl.

SMHI's wind resource maps have been made with the wind atlas program WAsP. The energy content has been calculated for a number of points in a grid. This means that the information is smoothed out. It is not detailed enough to make calculations for wind turbines at a specific site, but gives a general idea of the areas where the preconditions for wind power development are the best.

In 2006/07 the Swedish Energy Agency published a new wind resource map of Sweden, made by a meso-scale meteorological model, and later an updated version with a resolution of 500 × 500 m in 2011. Meso-scale models model the climate at large – air pressure, temperature, etc. – and with these meso-scale weather data wind resource maps are created. The Swedish maps were made with the MIUU-model developed at Uppsala University. Wind resource maps are nowadays available on the internet (see Figure 2.17).

Long-term correlation

If the wind speed is measured at a site for a shorter period, for example 6–12 months, the wind energy during a *normal* year can be calculated by using wind data from a nearby measuring mast for which long-term data are available, *if there is a correlation between the wind at the two sites*. The frequency distribution can be assumed to be the same.

By calculating the quotient of the measured average wind speed and the corresponding average wind speed from the meteorological station for the *same* period of time, this quotient can be multiplied by the long-term (minimum 5–10 years) average wind speed from the meteorological station. This normalized average wind speed is given the same frequency distribution as the winds at the meteorological station and after that the

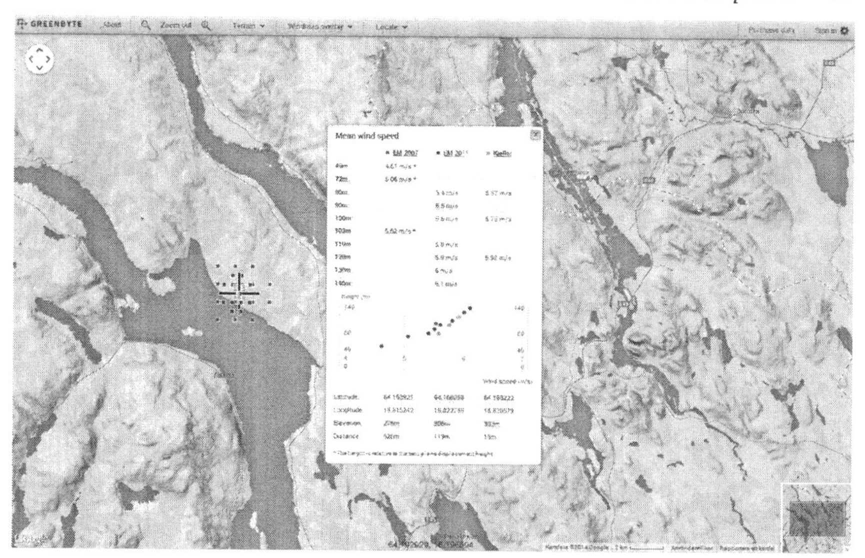

Figure 2.17 Wind resource maps. Uppsala University has made a new more detailed wind resource map for the Swedish Energy Agency. Data have been processed for seven different heights, 80, 90, 100, 110, 120, 130 and 140 m agl with a resolution 0.25 km^2. The company Greenbyte presents a wind resource map on the internet at www.windmap.se where smaller areas can be zoomed in on and the wind speeds on different heights are shown. The map shows the wind energy (average wind speed) at Kastlösa on Öland. Data from the Swedish MIUU maps from 2007 and 2011, and from Norwegian Kjeller are shown. The maps have been developed with mesoscale models

energy content of the wind at the wind turbine site can be calculated. This is the simple method of doing this; but there are other more advanced so-called MCP (measure-correlate-predict) methods, like the *long-term Weibull scale method*. Nowadays there are also databases with global wind data, like MERRA, based on measurements from satellites, which are used for long term correlation of local wind resources.

Since wind power started to develop, and turn into an industry manufacturing wind turbines to utilize this renewable energy source, the interest in understanding and mapping out wind resources has increased as well. The means and methods of evaluating these resources have improved each year, and many wind resource maps are nowadays available on the internet. It is very important to understand the nature and behaviour of the wind, and what impacts the land coverage, topography and local climate have on the moving air to be able to find and choose the best sites to utilize the power of the wind.

References

Bogren, J. et al. (1999) *Klimatologi, meteorologi*. Lund: Studentlitteratur.

DTU Wind Energy (n.d.). Accessed 2 December 2014 at http://www.vindenergi.dtu.dk/english

Gipe, P. (1995) *Wind Energy Comes of Age*. Chichester: John Wiley.

Jensen, N.O. (1986) *A Note on Wind Generator Interaction* RISØ-M-2411 Roskilde: Risoe National Laboratory.

Troen, I. and Petersen, L.E. (1989) *European Wind Atlas*. Roskilde: Risoe National Laboratory.

3 Wind turbines and wind power plants

The winds that move above our heads contain a lot of power. To be able to use this source of energy, this power has to be 'caught' and transformed to a form that can be used. This can be done by means of a turbine, which the wind causes to rotate, so that the turbine will turn an axis that can be connected to a millstone, a water pump or an electrical generator.

Conversion of wind energy

The wind drives the rotor on a windmill in the following fashion. The rotor blades are inclined in relation to the wind. The moving air pushes against the blades that start to move in one direction and the air moves in the other direction (action–reaction) (see Figure 3.1).

If the rotor is allowed to rotate without any load, it will accelerate, up to a limit. When the speed of rotation increases, the *apparent wind direction* will get closer to the direction of the blade (the chord). Finally it will become parallel to the blade direction. Then no power will impinge on the blade, and the speed of rotation will decrease (see Figure 3.2). The apparent wind direction will return to its previous state, the rotor will accelerate again, and this procedure will repeat itself.

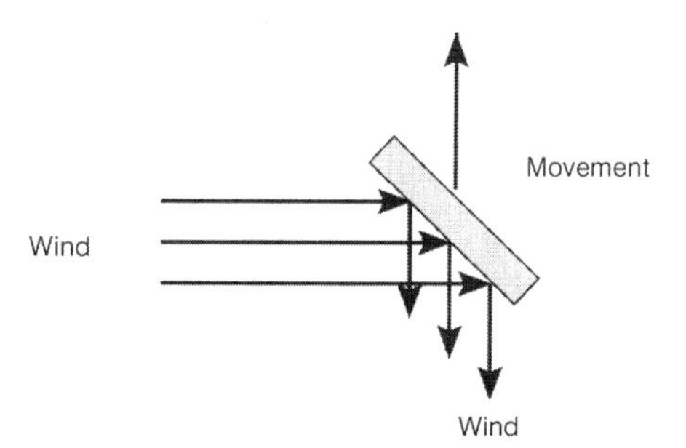

Figure 3.1 Wind against stationary blade. When the wind starts to blow on a rotor with blades that are stationary, the air is forced to move in one direction while the blade is pushed in the other, and the rotor will start to turn (illustration: Typoform)

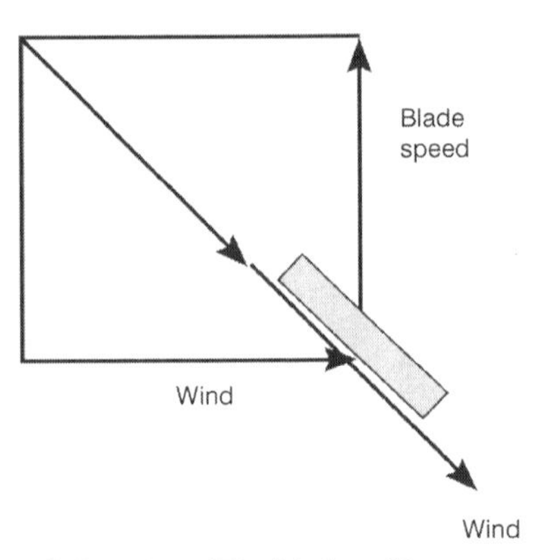

Figure 3.2 Apparent wind direction. The blade will react to the resulting wind that is the sum of the vectors from the horizontal wind speed and the speed of rotation (illustration: Typoform)

When the energy in the wind is converted, it is done with a special purpose: to drive a water pump, an electrical generator, a so-called *load*. The axle that is connected to a millstone, a generator or some other device that has to be turned around, offers resistance. There has to be a good balance between the load/work on one side, and the ability of the rotor to catch and convert the energy in the wind on the other. How should a wind turbine be constructed to be efficient?

Windmill engineers in medieval times, who were very clever and skilful, constructed efficient windmills, but their constructions were based solely on experience. In England in 1759, John Smeaton started to conduct some practical experiments. He was trying to find the most efficient blade angle for a windmill. He built a model of a rotor where the blade angle could be changed and mounted it on a horizontal pole on a device in a barn, so that the pole could be rotated.

He attached a wire to the rotor axis that was drawn over a wheel so that different weights could be attached to the other end of the wire. One of his farm hands was then assigned the task of setting the pole with the experimental rotor on the end in motion, by pulling a rope that was wound around a vertical axis, so that the experimental rotor moved around in a circle inside the barn. The wind speed that hit the rotor would be the same as the peripheral speed of the pole. By changing the angles of the rotor blades and observing which weights the rotor had the power to pull, he managed to find the most efficient blade angle.

The rope pulled around the small rotor at the end of the pole; weights were placed in the tray to measure the power of the rotor. Smeaton tested blades of different forms and angles. He found that the optimal rotor should have an angle of 18 degrees to the plane of rotation in the inner half of the blade, and be twisted to 16, 12 and 7 degrees in the three outer sixths of the blade. He developed a twisted blade.

The construction of modern wind turbines is based on much more advanced experiments and theories. There are many questions to tackle. How many blades should a rotor have? How large a part of the swept area should they cover? Which form should the blades have?

The wind in a stream tube

Wind is air in motion. Air has mass, and the power of wind is the product of the cube of the wind speed and mass of air that passes the rotor disc during a specified time. Energy can neither be created nor destroyed; it can only be converted from one form to another. To convert the kinetic energy of the wind the moving air has to be slowed down.

To convert all of the wind's kinetic energy, the moving air has to be retarded completely. That is, however, not feasible. When a rotor slows down the wind so much that it is standing still behind it, the immovable air will stop the airflow. If you mount a solid rotor it will stop the airflow and the air will escape outside its limits. The wind must be able to pass through the rotor, and also be able to move on behind the rotor. It has to keep some of its speed.

The rotor of a wind turbine is a free turbine: the wind and the turbine are unshrouded. In a hydropower station the water is led into a tube and a wall surrounds the turbine so that no water can escape, and thus it will be possible to utilize almost 100 per cent of the kinetic energy of the water stream. With a free turbine this is theoretically impossible.

Fluid mechanics is a specialist subject within physics that studies the properties of fluid matter, fluid liquids and gases. Aerodynamics is the part of fluid mechanics that is about air. A stream tube is an imaginary tube in the direction of the wind into which the turbine fits (see Box 3.1).

How efficiently the power in the wind is utilized depends on how much the wind is retarded by the rotor. If it is retarded too much, or too little, efficiency will be low. The cross-section area of the stream tube in front of the turbine, and thus the mass flow that will pass the turbine, will decrease when the wind speed is retarded (see Figures 3.4 and 3.5).

The modern theory for wind turbines was created by the German scientist Albert Betz from Göttingen and was further developed by Hans Glauert and G. Schmitz. Betz proved that a wind turbine is most efficient when the wind speed is retarded by one third just in front of the rotor, and by another third behind the rotor. The undisturbed wind v is retarded by the rotor to $2/3v$ and will decrease to $1/3v$ behind the rotor before it regains its original wind speed due to the influence of the surrounding wind. The power in the wind is then used most efficiently, in this case 16/27 (59 per cent) of the power in the wind can be extracted if aerodynamic and mechanical losses are ignored. The rotor of a wind turbine can at most utilize 59 per cent of the energy content of the wind, according to basic theory.

Box 3.1 Wind in a stream tube

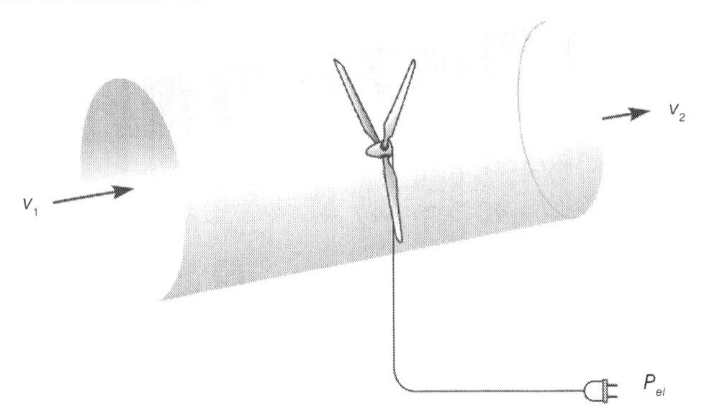

Figure 3.3 Wind in a stream tube

The power in a stream tube is $P = \frac{1}{2}\dot{m}v^2$. If a wind turbine is placed in the stream tube, a part of this power will be converted to electric power. However, the power that enters the tube equals the power that leaves the tube plus power that has been extracted by the turbine:

$$P_{\text{before}} = P_{\text{el}} + P_{\text{after}}$$

$$\frac{1}{2}\dot{m}v_1^2 = P_{\text{el}} + \frac{1}{2}\dot{m}v_2^2$$

$$P_{\text{el}} = \frac{1}{2}\dot{m}(v_1^2 - v_2^2)$$

The power that can be extracted by the turbine depends on how much the wind speed is retarded. P_{el} can be maximized by choosing a suitable value for the retardation of the undisturbed wind speed v_1.

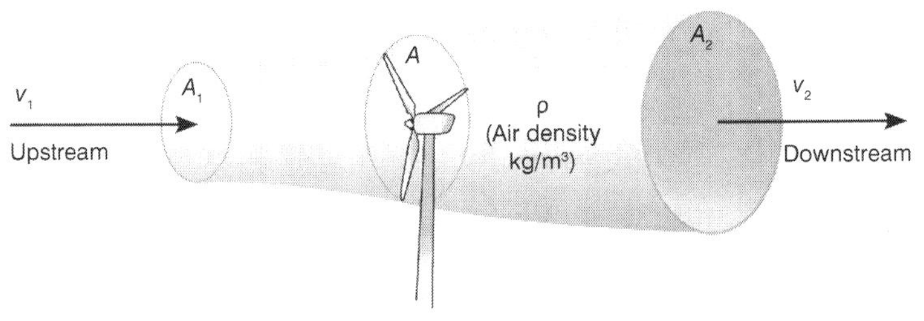

Figure 3.4 Retardation of wind speed. The same amount (kg per second) will pass the areas A_1, A and A_2 (otherwise air would accumulate in the tube). The mass flow ($\dot{m} = Av\rho$) is the same. This means that $A_1v_1\rho = Av\rho = A_2v_2\rho$. Since wind speed is retarded ($v_1 > v > v_2$), $A_1 < A < A_2$, the stream tube expands. (illustration: Typoform)

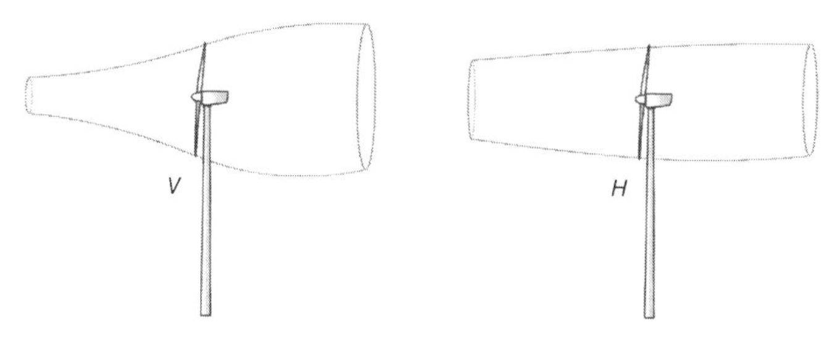

Figure 3.5 Optimal retardation. In the example to the left (*V*) retardation is very strong. The share of extracted power is large, but it is taken from a very narrow stream tube, so that the total extracted power will be small. In the example to the right (*H*) a small share is extracted but from a very wide and thus powerful stream tube. In this case the total extracted power will also be small. The optimal retardation is somewhere between these extreme cases (illustration: Typoform)

Alfred Betz showed that half of the retardation happens at the turbine and the rest behind the turbine after the wind has passed through it. The change of the wind speed does not happen stepwise but continuously. The undisturbed wind speed v is reduced to $v(1-a)$ when it passes through the turbine and to $v(1-2a)$ some distance behind it (the retardation starts about one rotor diameter in front of the turbine and reaches its maximum about one diameter behind it). The unit a is called *interference factor*. If this factor is 0.5, the wind speed behind the rotor will be reduced to zero. This means that $0<a<0.5$. The theoretical maximum share of the power in the undisturbed wind that can be utilized is 16/27 – that corresponds to 59.3 %. This maximum is reached for $a = 1/3$ which means that the turbine retards the wind speed by 1/3 at the turbine and with another 1/3 behind the turbine.

The share of the power in the wind that can be utilized by the rotor is called the *power coefficient*, C_p. Its maximum value is $Cp_{max} = 16/27$ (≈ 0.593). For real turbines C_p is lower, due to aerodynamic and mechanical losses, and the value also varies for different wind speeds. The power that a wind turbine can attain can be expressed as

$$P = \frac{1}{2}\rho A v^3 C_p$$

How should a wind turbine rotor be constructed? Most old windmills have four rectangular blades that cover about 20 per cent of the swept area. This was mainly due to practical considerations; it was easy to build and worked well. Windmills for water pumping that were developed in the United States in the nineteenth century had blades that covered almost the whole swept area, and the wind passed through slots between the blades. Modern wind turbines use three slender blades that cover not more than 3–4 per cent of the swept area. They are however much more efficient than their predecessors.

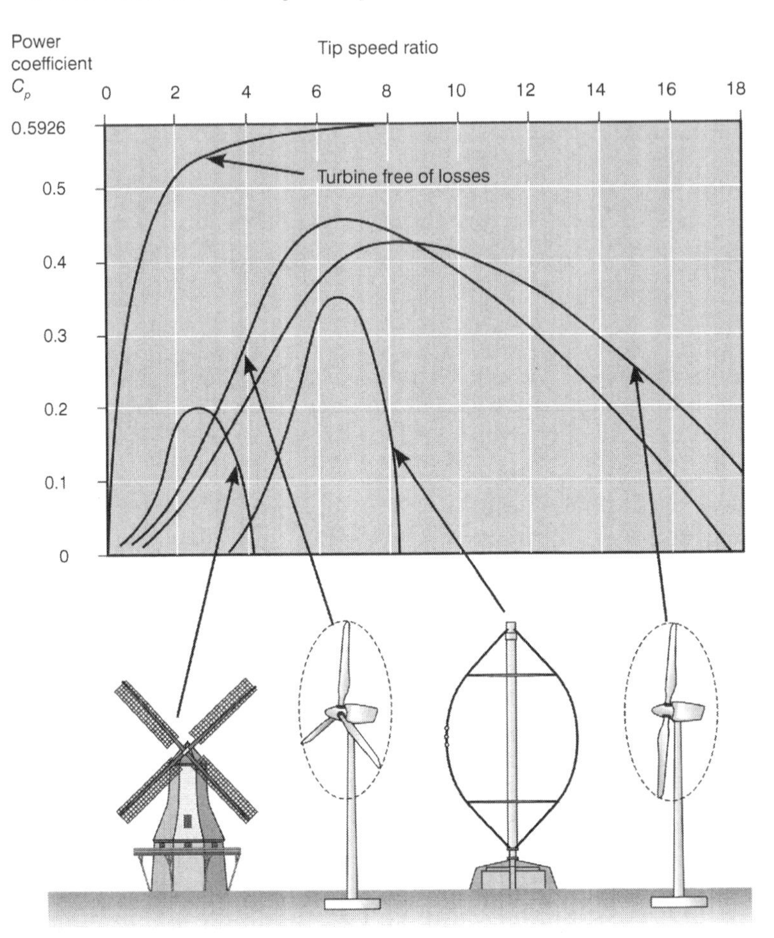

Figure 3.6 Tip speed ratio diagrams. The tip speed ratio is the relation between the tip speed v_{tip} and the undisturbed wind speed v_0 and is signified by λ: $\lambda = v_{tip}/v_0$. The power coefficient C_p gives a measure of how large a proportion of the power in the wind a turbine can utilize. The theoretical maximum value of C_p is $16/27 \approx 0.5926$. The diagram shows the relation between tip speed ratio and power coefficient for different types of wind turbines: a) windmill, b) modern turbine with three blades, c) vertical axis Darrieus turbine, d) modern turbine with two blades (illustration: Typoform, based on Södergård, 1990)

The optimal rotor has, theoretically, an infinite number of infinitely narrow blades, but you can't manufacture such a turbine. With blade element theory it is possible to calculate the optimum total blade width that can then be divided by the desired number of blades. The fewer blades that are used, the higher rotational speed is needed to get the same efficiency.

The rotational speed of the rotor in relation to the undisturbed wind speed plays a crucial role for the efficiency of the turbine. This is called the *tip speed ratio*. It is a measure of the relation between the tip speed (the speed of the tip of the rotor blades) and the undisturbed wind speed (before it has been retarded by the rotor). From practical experience and by theoretical calculation, the optimal tip speed ratio can be calculated for different types of rotors (see Figure 3.6).

Figure 3.6 shows that a windmill has a narrow range of tip speed ratios. It is most efficient for a ratio of 2, and the same applies for other older types of wind turbines. Modern turbines with two or three blades have a maximum at a ratio of 10 and 7 respectively. They have a broad span; they are quite efficient over a wide range of tip speed ratios.

If the ratio is 1 the blade tip has the same speed as the wind. The apparent wind direction will be 45 degrees (to the plane of rotation), but since the wind is retarded by 1/3 before it reaches the rotor, the apparent wind direction will be 34 degrees. To be able to utilize the power of the wind, the angle of the blades has to be smaller than that, half as big: 17 degrees. Windmills, with a tip speed ratio around 2, often have a blade angle (to the plane of rotation) of 15 degrees.

The blade tip of a modern wind turbine has a speed ten times as fast as the wind speed. It may seem implausible for a wind of 5 m/s to drive a blade with a speed of 50 m/s. This is however what actually happens and it is also possible to explain how this is possible.

Aerodynamic lift

If we look at other devices that are driven by the wind, like a sailing ship or an ice-yacht (where the phenomenon is more obvious due to lower friction, it will be easier to understand how this aerodynamic lift works. When a yacht sails in the wind direction, with the wind coming from the rear, it can't move faster than the wind. That is quite obvious. The wind pushes the yacht from behind.

If the yacht beats to windward, that is, sails in a direction where the wind comes diagonally from the front, the wind can't push the yacht. Instead the yacht is *pulled* by the difference in pressure that is created when the wind passes the sail (that has a shape similar to an airplane wing). With a sailing yacht it is not very obvious that it will move faster than the wind, since water creates strong friction. An ice yacht can however easily reach speeds of 100 km/hour with a wind of 8 m/s (30 km/hour). The yacht in fact moves three times faster than the wind. It is moved by *lift*, just like an airplane is moved upwards and kept in the air by aerodynamic lift.

If you stretch your arm out through the window of a car that moves with a good speed, you can feel your arm pushed backwards. If you hold the arm straight with your hand parallel to the road, and change the angle slightly, you can suddenly feel that it is drawn upwards. The hand and arm work like the wing of an airplane, and with the right angle you can feel a strong *lift force*.

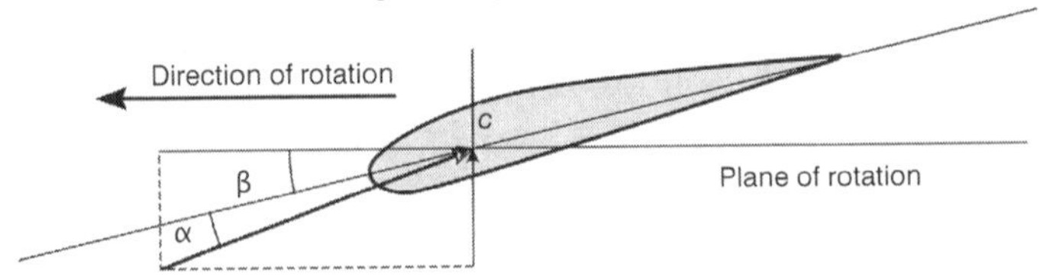

Figure 3.7 Airfoil. An airfoil is divided by a centreline, the *chord*. The *angle of attack* is defined as the angle between the chord line and the direction of the apparent wind. To get a suitable angle of attack the blades of a wind turbine rotor are set at an angle to the plane of rotation, the *blade angle*. This blade angle is signified by β. The airfoil also has a centre point, *c*, that is situated on the chord about 25 per cent from the front edge of the airfoil (the distance depends on the form of airfoil). The sums of lift and other forces are summarized and applied to this point (illustration: Typoform)

These two forces determine the properties of an aerodynamic blade profile (or airfoil). One force that pulls backward, *drag* (*D*), another force that pulls upward, *lift* (*L*). It is this lift force that makes airplanes fly and the rotor of modern wind turbines rotate. (The lift force is actually present also on so-called drag devices, like wind mills and Savonius rotors, but to a lesser degree, otherwise the tip speed ratio would not pass 1.)

The properties of an airplane wing are defined by its airfoil. It is the form of the airfoil, in combination with the angle of attack, which determines the lift and other properties. These properties are detected by tests in a wind tunnel (see Figure 3.7).

When a stream of air hits the leading edge of an airfoil, some of the air passes over it and some below. When the air stream passes the airfoil a lift force is created, that depends on the angle of the wind direction in relation to the chord: the *angle of attack* α (see Figure 3.8).

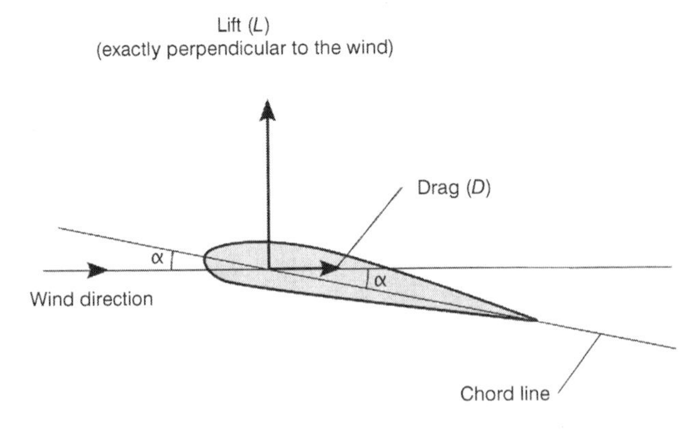

Figure 3.8 Lift. Lift *L* and drag *D* are functions of the angle of attack α. The force *L* is always perpendicular to the direction of the apparent wind, while the force *D* is applied in the direction of the apparent wind, and perpendicular to *L* (illustration: Typoform)

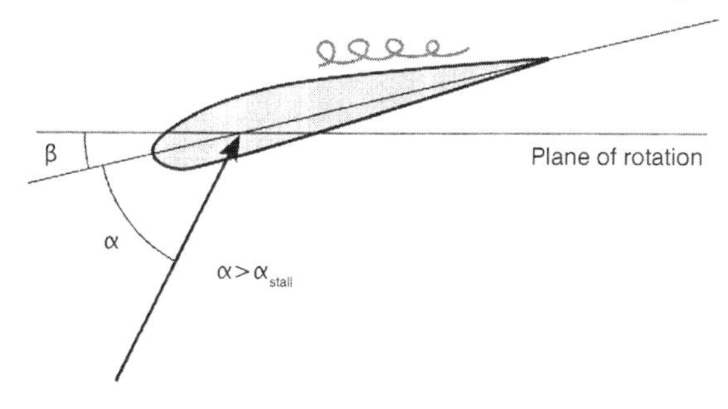

Figure 3.9 Stall. When the angle of attack α becomes too large, the airflow cannot follow the topside of the airfoil all the way to the trailing edge. Eddies are created which reduce lift and increase drag. For a wind turbine with constant rotational speed and fixed blade angles, the angle of attack will increase when the wind speed increases; the angle of the apparent wind direction will increase. Therefore stall can be utilized to reduce lift, and thereby the power, when the wind speed becomes larger than a specified speed (illustration: Typoform)

If the angle of attack is too large the stream of air that passes on the top side of the airfoil cannot attach to the profile all the way to the back edge (called the trailing edge). The flow will stall (eddies are created). When that happens, the lift force will decrease. This property is used by stall-regulated turbines to limit the power of the rotor (see Figure 3.9).

The strength of the lift force also depends on the form and width of the airfoil and the wind speed. Modern wind turbines utilize these aerodynamic properties to optimize the airfoils that are used on the rotor blades of the turbines.

Types of wind turbines

There are several different design concepts for wind turbines. One basic classification is horizontal axis wind turbines (HAWT) versus vertical axis wind turbines (VAWT). Horizontal axis wind turbines can have the rotor *upwind*, that is facing the wind or *downwind* so that the wind will pass the tower and nacelle before it hits the rotor (see Figure 3.10). Today most turbines have an upwind rotor, but there are turbines, from prototypes in the MW class to smaller turbines with a nominal power of 20–150 kW as well as water pumping wind wheels from the nineteenth century, with downwind rotors.

Horizontal axis turbines

All these types of wind turbines have been built and used in practice. The windmill and the wind wheel have a long history. The windmill has played its part, and there are around one million wind wheels used for water pumping in use around the world. They have a very robust design, with quite simple

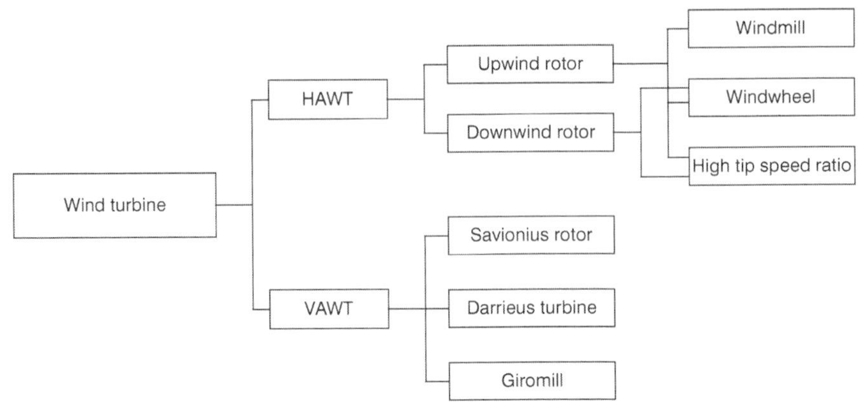

Figure 3.10 Types of wind turbines

components that are easy to maintain and repair. The advantage with a wind wheel compared to a turbine with few slender blades and high rotational speed is that it starts more easily, since the blades cover a much larger share of the swept area. This is an advantage for a water pump, since it takes a lot of power to get it running.

Turbines with a high tip speed ratio were first used as battery chargers. Today they are used to produce electric power that is fed into the power grid. There are small micro turbines (with 2–6 blades) for battery charging.

In the 1980s there were many different designs of grid-connected wind turbines, with either two or three rotor blades, some with the rotor downwind and others with the rotor upwind. The advantage with a downwind rotor is that it automatically adjusts itself to the wind direction; however, with sudden changes in wind direction this does not happen. Since the beginning of the twenty-first century, turbines with three-bladed upwind rotors completely dominate the market (see Figure 3.11).

Vertical axis turbines

The advantage with a vertical axis wind turbine is that the generator and gearbox can be installed at ground level, so they are easy to service and repair. Both the Savonius rotor and the Darrieus turbine are manufactured commercially, but as small models that are used for different niche applications, like battery charging in areas without a power grid (see Figures 3.12–3.15).

The Finnish engineer and inventor Georg Savonius developed a vertical axis wind turbine in 1924, which is called the Savonius rotor. It consists of a vertical S-shaped surface that rotates around a central axis. By positioning the two halves of the rotor so that they overlap, and the wind can slip through the middle, efficiency can be increased. Nowadays Savonius rotors are most commonly seen as advertisement posters in front of restaurants and

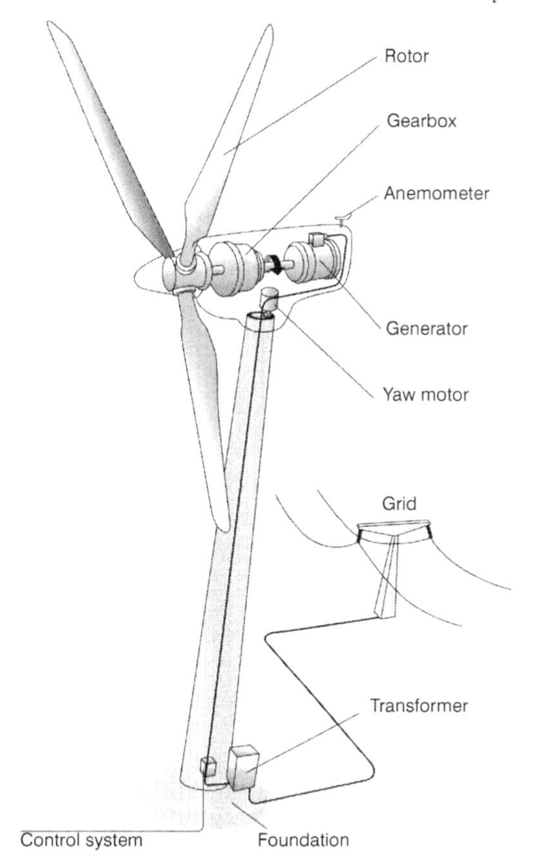

Figure 3.11 The main components of a wind turbine. A wind turbine consists of the following main components: foundation, tower, nacelle (generator, gearbox, yaw motor, etc.), rotor, control system and transformer. The turbines that dominate the market today have high tip speed ratios, where the blade tip moves five to seven times faster than the wind, a rotor with three blades and a rotational speed of 5–30 revolutions per minute. Most manufacturers offer several options of their models, with different hub heights and/or rotor diameter, so that the turbines can be tailored for specific sites.

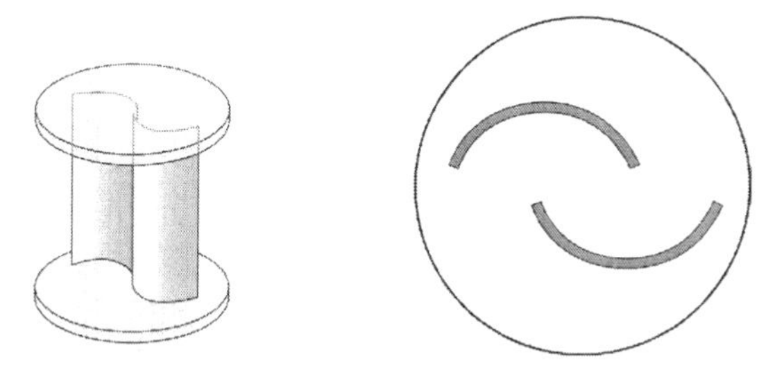

Figure 3.12 Savonius rotor (illustration: Typoform based on Gipe, 1993)

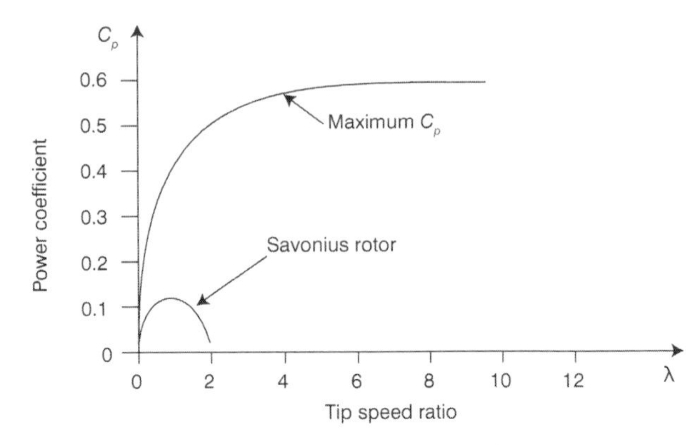

Figure 3.13 Efficiency of the Savonius rotor (source: Claesson, 1989)

shops. They revolve, but produce no power. There are, however, power-producing Savonius rotors as well. They are easy to maintain and reliable, but need much material in relation to the power produced and are not very efficient (see Figure 3.13).

The Savonius rotor is used as a battery charger on lighthouses and telecom masts. It can also be used as starter motor for a Darrieus turbine (see Figure 3.14). The French engineer Georges Darrieus invented this eggbeater-shaped wind turbine in 1925. It can have two to four blades, which form bows from the top of the tower to the machinery that is sited at ground level. The blades are symmetric and very thin. The form directs the centrifugal force to the points where they are connected to the central axis, so that the

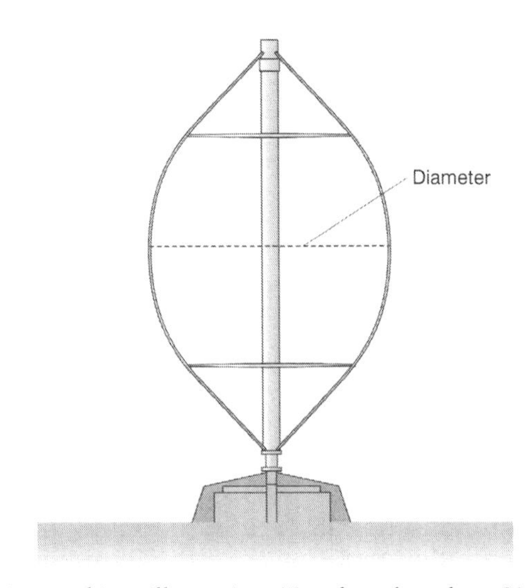

Figure 3.14 Darrieus turbine (illustration: Typoform based on Gipe, 1993)

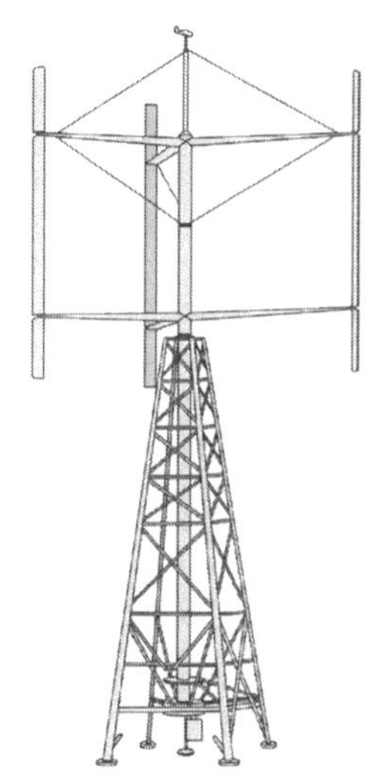

Figure 3.15 Giromill (illustration: Typoform based on Gipe, 1993)

bending moments are minimized. In most materials the tensile strength is stronger than the bending strength. A Darrieus turbine does not need much material, relative to the power produced.

How can such a symmetrical turbine start to revolve? Well, it can't, it needs a starting motor. But as soon as it starts to revolve, the wind will immediately take over, since the wind and the revolution speed together resulting in a wind that creates lift in the direction of revolution.

The swept area on a Darrieus turbine is $A=2/3D^2$. It has a narrow range of tip speed ratios of around 6 and a power coefficient just above 0.3. Several Darrieus turbines have been built, from a large MW prototype in Canada, commercial turbines in the 150 kW range and small models of a few kW.

A giromill is a turbine with two or more straight vertical blades that form an 'H' which are connected to a vertical axis (see Figure 3.15).

A mechanical device that changes the blade angle during rotation can increase its efficiency. This makes it self-starting, but the blade angles have to be adjusted in relation to the wind direction to start. The swept area of a giromill is the height times the diameter: H × D. The strong load on the blades' points of attachment and the bending moments are weak points of this design.

A giromill is more efficient than a Darrieus turbine and has a wider range of tip speed ratios but it is not as efficient as a horizontal axis turbine with high tip speed ratio.

The wind turbine rotor

A wind turbine rotor consists of rotor blades connected to a hub where the blades are mounted. Most commercial wind turbines have a three-bladed

Figure 3.16 Wind turbine with two rotor blades. Wind turbine NWP 1000 from the Swedish company Nordic Windpower has a two-bladed rotor with a teetering hub. With such a 'soft' design concept the whole turbine can become slimmer and cheaper (photo: Tore Wizelius)

Figure 3.17 Wind turbine with one rotor blade. This turbine from the Italian manufacturer Riva Calzoni has a one-blade downwind rotor. The blade, that is balanced by a counterweight, is flexible in the hub and moves backwards when the wind speed increases, reducing the load on the turbine (photo: Riva Calzoni)

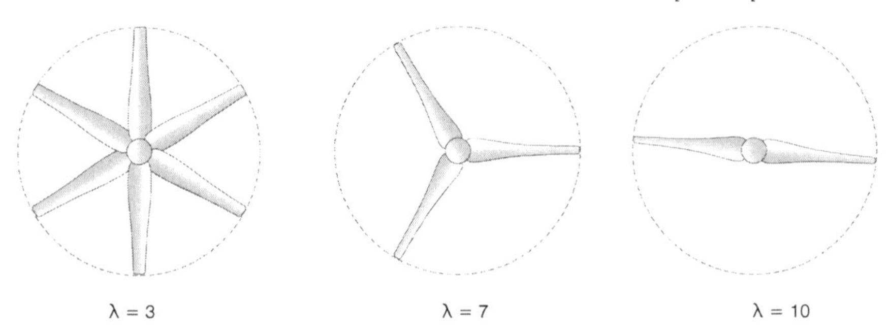

Figure 3.18 Relation between number of blades and tip speed ratio. Wind turbines with six, three and two blades respectively and their tip speed ratios (illustration: Typoform based on Södergård, 1990)

rotor. There are, however, turbines with two blades and, in fact, there are also turbines with only one single blade. The advantage of fewer blades is that the weight of the rotor and of many other components of the turbine decreases. The proportion of the power in the wind that can be converted decreases with fewer blades, but from an efficiency point of view, the differences are negligible or, at least, easy to compensate for by increasing the length of the rotor blade a little.

On three-bladed turbines the connection between hub and blades is rigid. On a turbine with two blades or one single blade, the blades can be mounted so that they are flexible in the vertical plane. On a so-called teetering hub, the two blades can teeter a few degrees across the hub, which reduces loads on the turbine (see Figures 3.16 and 3.17).

The tip speed ratio of the rotor

To utilize the power in a wind-efficient way, the rotor has to have suitable rotational speed relative to its size (i.e. the rotor diameter) and the wind speed. It has to have an efficient tip speed ratio. The tip speed ratio of a turbine depends on the number of blades; when the number of blades decreases, the tip speed ratio should increase. This means that for turbines with the same rotor diameter, a one-bladed turbine needs a higher rotational speed than a two-bladed turbine which in turn needs a higher rotational speed than a three-bladed turbine, and so on (see Figure 3.18).

There is also a relation between tip speed ratio, rotational speed and the size of the rotor. The tip speed ratio is the relation between the speed at the tip of the rotor blade and the undisturbed wind speed: $\lambda = v_{tip} / v_0$. At a given rotational speed the tip speed increases with the length of the blade.

The rotational speed is usually signified by n and with the unit revolutions per minute (rpm). The tip speed is given in the same unit as the wind speed, metres per second (m/s). The tip speed depends, beside the rotational speed, also on the radius of the rotor, and is calculated with this formula:

$$v_{tip} = \frac{n2\pi R}{60} \text{ m/s}$$

where R is the radius of the rotor.

The tip speed for a rotor with a 10-metre radius and a revolution speed of 30 rpm is

$$v_{tip} = \frac{30 \times 2\pi \times 10}{60} \approx 30 \text{ m/s}$$

If the radius is increased to 20 m the tip speed will increase to ≈ 60 m/s. To keep the tip speed at 30 m/s when the rotor radius is increased to 20 m the rotational speed can be reduced to 15 rpm.

When the blades of a wind turbine rotate, the speed at the tip of the blade is higher than at the root or in the middle part of the blade. For a wind turbine with a 20-metre radius and a rotational speed of 30 rpm the tip speed is 60 m/s but the speed at the middle of the blade is only 30 m/s.

Since the speed of the blade segments increases from the root to the tip, the apparent wind direction will change as well; from root to tip, the apparent wind direction will move towards the vertical plane. To get the same angle of attack along the whole blade, it has to be twisted (see Box 3.2).

Lift and circumferential force

The properties of the rotor also depend on the airfoil used on the blades. In the early days of wind power (in the 1970s and 1980s) airfoils developed for airplanes were used, mainly so-called NACA airfoils. From the 1990s airfoils developed specially for wind turbines have been used. One profile can have several different thicknesses. The last two digits in the type number of an airfoil give the relative thickness (thickness in relation to width) in per cent: the maximum thickness of NACA4412 is 12 per cent of the width.

On a wind turbine the lift from the rotor blades is utilized to make it revolve. The circumferential force is not the same as the lift force. The lift is always applied perpendicular to the apparent wind direction. The blade has a certain (constant or variable) angle to the plane of rotation. The airfoil also has some friction, which is called drag (*D*) that is applied in the apparent wind direction. From these forces we get a circumferential force F_{circ} in the plane of rotation, and a force perpendicular to the plane of rotation, the thrust F_{thrust}. The circumferential force that propels the rotor seems quite small in relation to F_{thrust}, but the power is large since the rotational speed is very high (see Figure 3.20).

The airfoils of a wind turbine should create a strong circumferential force and also have other properties suitable for wind turbines. These properties can be read from the graphs in Figure 3.21 which show the test results of profiles in a wind tunnel. The first shows the relation between the angle of attack and lift (*L*), the second, a gliding ratio diagram, shows the relation between the lift coefficient C_L and the drag coefficient C_D.

Box 3.2 Blade twist

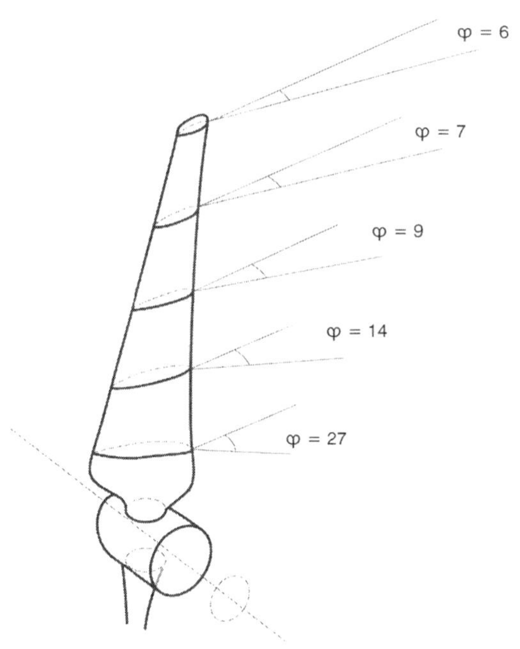

Figure 3.19 Apparent wind direction along a rotor blade with a wind speed of 9 m/s

v_o = 9 m/s

$2/3v_o$ = 6 m/s (the undisturbed wind will decrease to 2/3 just in front of the rotor disc).

$$v_{tip} = 60 \text{ m/s} \quad \varphi = 6$$
$$v_{0.8R} = 48 \text{ m/s} \quad \varphi = 7$$
$$v_{0.6R} = 36 \text{ m/s} \quad \varphi = 19$$
$$v_{0.4R} = 24 \text{ m/s} \quad \varphi = 14$$
$$v_{0.2R} = 12 \text{ m/s} \quad \varphi = 27$$

$\varphi = \alpha + \beta$

where ϕ = the angle of the apparent wind to the vertical plane

α = angle of attack

β = blade angle (pitch)

By twisting the blade (the blade chord's angle to the plane of rotation β) so that the angle decreases towards the tip, the angle of attack α can be kept constant (for a given wind speed): $\beta = \varphi - \alpha$

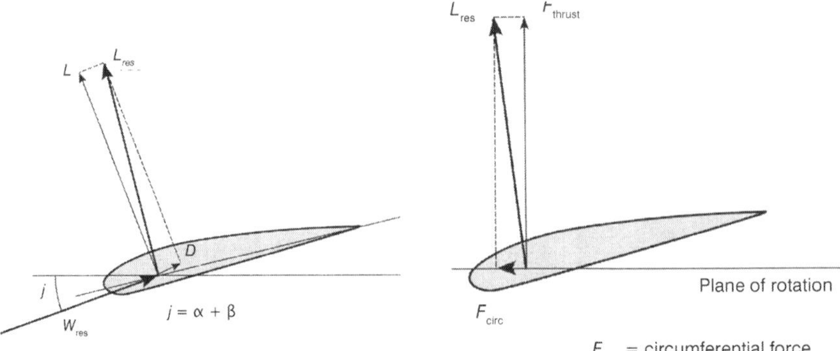

Figure 3.20 Lift and circumferential force. When a stream of air passes an airfoil drag D is created in the apparent wind direction (v_{res}) and lift L perpendicular to this. These two forces have a resultant lift: L_{res} (left diagram). L_{res} is divided into two forces, F_{circ} which is applied in the plane of rotation and makes the rotor revolve (i.e. useful force) and F_{thrust} (useless force) that is applied perpendicular to the plane of rotation. This force is absorbed by the main shaft, main bearing and tower (illustrations: Typoform based on Claesson, 1989)

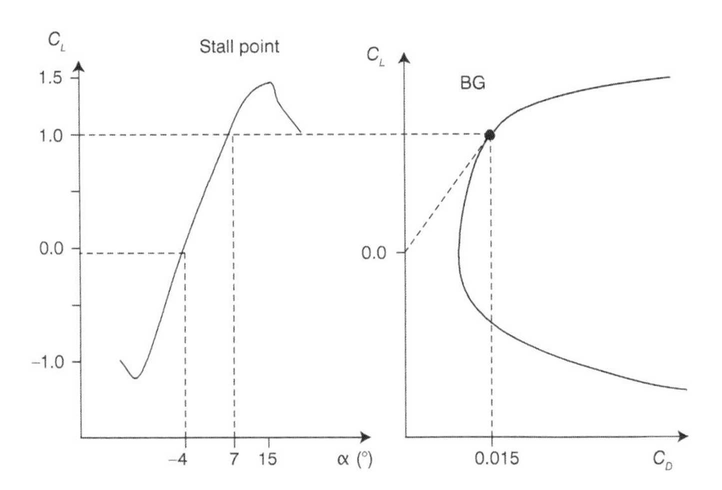

Figure 3.21 Airfoil diagrams. The left-hand graph shows the relation between lift and angle of attack. The airfoil that the diagram describes starts to have lift by an angle of attack of –4°; it reaches maximum at 15° and after that it will decrease. With such large angle of attack the air stream cannot stick to the surface of the airfoil, so that turbulence is created, the airflow separates and the airfoil begins to stall. The right-hand graph shows the relation between C_L and C_D. C_L (the lift coefficient) and C_D (the drag coefficient) are standardized aerodynamic coefficients that make it possible to calculate lift and drag for airfoils of different sizes. The best gliding ratio is found at the tangential point of a line from 0 on the C_L axis and the graph. The airfoil is most efficient with an angle of attack of 7° and has a safe distance to the point where stall occurs at $\alpha > 15°$

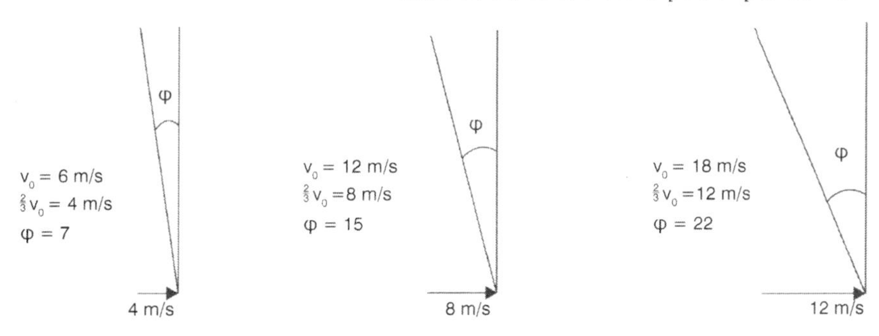

Figure 3.22 Apparent wind direction and wind speed. Apparent wind direction over a segment of a rotor blade with the rotational speed of 30 m/s. On wind turbines with constant rotational speed and blade angle, the angle between the plane of rotation and the apparent wind speed will increase when the wind speed increases. The angle of attack will increase simultaneously

The direction of the wind that passes the blades will vary along the blade, the angle in relation to the plane of rotation will decrease towards the blade tip. The apparent wind direction will also change each time the undisturbed wind speed changes. If the blade angle and rotational speed are constant, the angle of attack will change continuously and therefore also the lift, drag and gliding ratio on different parts of the blade (see Figure 3.22).

Up to the end of the 1990s most wind turbines had a fixed rotational speed, which for turbines with 1 MW nominal power and about 50 m rotor diameter would be around 25 rpm. Some turbines use two different fixed rotational speeds; one for low wind speeds, when a small generator is used, and a higher for stronger winds when the large generator cuts in. In this way the rotor can operate close to the optimum tip speed ratio at all wind speeds. In the ideal case, however, it is the tip speed ratio that should be fixed, and not the rotational speed. The rotational speed should increase with the wind speed. However, if the rotational speed of the rotor varies, the generator will also get a variable speed, so that the voltage and frequency will vary as well.

A few manufacturers in the 1980s used variable rotational speed, combined with power electronics (rectifier and inverter) but in recent years more and more manufacturers have changed from fixed to variable rotational speed designs. One driving factor behind this has been that the price of power electronics has fallen quite quickly. Another is grid operators' demand for power quality from wind turbines.

Power control

The wind that makes the rotor revolve also pushes the whole turbine backwards. If this thrust force gets too strong, the components in the turbine will be overloaded, so that they break or the whole turbine can be pushed over. Strong winds can break the trunks of trees, or lift a roof off a house. Wind turbines stand where they have been installed, and have to endure the worst weather conditions that can occur at the site where the turbine is located.

Power control means that the turbine limits the proportion of the available power in the wind when the wind speed reaches a preset value – the rated wind speed. This is usually set at a value of between 12 and 16 m/s. The rated wind speed differs between models and manufacturers. It can also be adapted to the wind conditions at a specific site, by altering the rotor diameter to the generator's rated power. When the wind speed increases to storm force the rotor will be stopped and parked to protect the turbine from damage. This *cut-out wind speed* is set to about 25 m/s for most turbines.

There are two different methods of power control for large wind turbines: *pitch* and *stall* control. Turbines with pitch control have rotor blades that can be turned on their longitude axis from the hub. By turning the blades the angle of attack can be adapted to the wind speed. When the wind speed increases above the rated wind speed, the blades are turned so that the power that is extracted from the wind is reduced and kept constant at the rated power. Thus, the power curve will level out. When the cut-out wind speed is reached, the blades are turned out of wind – *feathered* – so that the wind can blow through the rotor without creating any lift, and the rotor stops rotating.

Turbines with *stall control* have a blade profile that creates eddies (turbulence) on the upper side of the blade when the wind speed increases over the nominal wind speed. The lift will decrease and the drag will increase. In this way the power that is extracted from the wind can be kept close to the nominal power of the turbine even when the wind speed is higher than the rated wind speed. It is difficult to manufacture a rotor blade that will stall exactly as much as desired at a specific wind speed to keep the power constant at the rated power in strong winds. This is, in fact, impossible because of variations in the air density with temperature and air pressure.

The stall usually increases gradually. It starts at 8–9 m/s and increases so that the power is kept as close as possible to the rated power of the turbine when the wind speed increases above the rated wind speed. When the cut-out wind speed is reached, the aerodynamic brakes will be activated to stop the rotor. This brake is set at the blade tips in most cases and the tip section of the blades will be twisted to stop the rotation.

Stall control has been developed into so-called *active stall*; the rotor blades can be turned along the longitude axis, just like pitch-controlled turbines. Pitch-controlled turbines turn the blades so that the wind will pass through the turbines more easily by decreasing the angle of attack to reduce the extraction of power when the wind speed increases. Turbines with active stall control, however, turn the blades the other way, increasing the angle of attack, and thus inducing a stall.

Box 3.3 Power control

The windmills of the past were run manually by the miller. When winds were too strong, the windmill was not used. When they were out of use, the rotor blades did not have sails (they were made of sailcloth or wood boards), and the rotor was parked with a chain. If unexpected strong winds occurred when the windmills were running, some could reduce the coverage of the rotor while running, by a safety device which would strip off the sails from the rotor. Others had self-regulating sails with shutters, so-called patent sails. Most windmills, however, had to be stopped with mechanical brakes or turned out of wind manually. This was not easy, and many windmills caught fire from the sparks from the brakes when the rotor got out of control. That is why there are so few windmills left today, most of them have vanished into smoke.

Wind wheels had other methods for power control. A large oblique vane was a simple method. When wind speed increased, the pressure from the wind on the vane increased so much that the rotor was turned out of the wind. On the Excenter turbine the nacelle was mounted eccentrically on the tower, so that it started to turn out of the wind when the wind pressure on the rotor exceeded the rated wind speed. On other small turbines, the rotor and nacelle are turned over backwards by a hinge in the back of the tower, so that the rotor assumes a helicopter position. The power decreases when the rotor swept area is reduced.

The rotor blades can also be used for power control, either by using an airfoil that stalls when the wind speed gets high enough, or by making the blades turn along the longitude axis. Many more or less complicated methods for this have been developed. A centrifugal regulator consists of a metal bar with a weight, attached to the blade close to the hub. The blades can turn, but are held in position by a spring. When the rotor revolves, the weights apply a force created by the centrifugal power to the blades. This force increases with the rotational speed and when the force gets stronger than the force of the springs, the blades start to turn and reduce the power extracted from the wind.

Wind turbines with variable rotational speed don't need to regulate the blade angle, since rotational speed increases with the wind speed to keep the angle of attack constant (and optimized). When a certain rotational speed and power output has been reached, this kind of turbine also has to control the power by pitching the rotor blades.

Rotor blade design

In practice, rotor blades must have many different properties and have to be designed to stand the strong loads that they will face. Therefore it is not suitable or possible to use ideal blades that give maximum efficiency. Most blades are designed with three different airfoils: a very thick one close to the hub, an airfoil with average thickness in the middle part of the blade, and a very thin one close to the tip. The part closest to the hub has to be thick and strong to make a solid connection of the blade to the hub. This part of the rotor blade is very thick and almost round, like a pipe. This innermost part of the rotor has a small area and does not contribute much to the power production.

The outer part of the blade often has a less than the optimum blade width, since more is gained by increasing the swept area than is lost by using a less efficient blade width that needs more material. Weight is the decisive factor for how long the blades can be.

Stall-controlled turbines with a fixed blade angle would be very hard to start in slow winds if the blades are twisted so as to attain optimum efficiency at a constant rotational speed. Part of the blade is therefore designed with a blade angle that will give enough lift to enable the rotor to start revolving at a wind speed of 3–5 m/s.

When the wind speed approaches the rated wind speed on a stall-controlled turbine, and power has to be controlled by stall, the revolution would become jerky if the wind stalled along the whole blade at the same time. This can be avoided by using a twist and a combination of airfoils so that the stall varies on different segments of the blade. The stall starts close to the root of the blade and spreads along the blade when the wind speed increases so that the power will level out when the rated power/wind speed is reached.

Pitch-controlled turbines can turn their blades to get the rotor to start to revolve after calm conditions. They can also vary the blade angle continuously to adapt it to the prevailing wind speed, and very large turbines can change the blade angle during each rotation to adapt the angle to the wind speeds at the different heights the blade moves through during each rotation. This has, however, not been a reasonable strategy, because wind speeds change so quickly and frequently. It would be difficult to regulate the blades fast enough and components would wear out quickly.

Pitch control is used mainly to control the power when the wind speed is above the rated wind speed. Pitch-controlled turbines can use different airfoils than those just described, as they don't depend on stall. Turbines with variable speeds can keep the blade angle constant since the angle of attack will be constant too.

Several manufacturers of stall-controlled turbines have developed so-called *active stall*. The advantage of this is that the power can be better controlled. The power in the wind depends not only on wind speed, but also on air density. It varies with air pressure and temperature. In conditions with the same wind speed, the wind will contain more power in winter when temperature is –10° C than in summer when it is plus 20°C. In winter a stall-controlled turbine with nominal power of 1 MW can produce 1.1 MW, which is not good for the generator. In summer the opposite applies, so that the turbine can't utilize its full capacity. By regulating the blade angle, the turbine can be adjusted to optimize the turbine in all weather conditions. The power can also be controlled when the wind speed is higher than the rated wind speed so that the power output stays at the rated power level.

Wind turbines with pitch control have a pitch mechanism installed in the hub. Some manufacturers use a hydraulic system for this; with a pump in the nacelle and a piston that passes through the main shaft to the hub, the movement is transferred mechanically to the blades. Other manufacturers use an electric system, with an electric motor connected to each of the blades. The latest development is to have pitch control on individual blades so that they can be turned independently of each other. On very large turbines the hydraulic system can be installed in the hub instead of in the nacelle.

If a rotor runs without a load, it will accelerate very quickly to its maximum tip speed ratio. A three-bladed turbine is most efficient at a tip speed ratio of 6–7, but without a load the tip speed ratio rapidly increases to 18 before it loses its power. The tip of a rotor blade on a turbine with 50 m rotor diameter will then have a tip speed of 180 m/s (650 km/hour) with a wind speed of 10 m/s; this is usually more than any blade will manage.

The load on a turbine that generates electric power is in the generator. If the turbine is disconnected from the grid, during a black out for example, the load disappears and the rotor runs free. Therefore all turbines must have an aerodynamic brake. If the blades can be turned, they will simply be turned so that the rotor stops. When the blades are turned to horizontal position, this is called 'feathering' the blades; they will lose lift and also be stopped by the friction from the air. Stall-controlled turbines have blade tips that can be twisted. The outer part of the blades is turned into a perpendicular position in relation to the rest of the blade. Inside the nacelle there is also a mechanical brake that can be applied when the rotor speed has decreased, and that can be used as an emergency brake if the aerodynamic brakes fail.

Rotor blades are exposed to great stress and strain. Since the wind is always changing, blades are exposed to continual load changes and have to be manufactured with material that can stand such stress and avoid fatigue. Steel and aluminium can't do this cost-effectively and most rotor blades are manufactured from glass fibre or epoxy. Wood is also a material that has good fatigue properties and there are blades made of laminated wood with a plastic coating. Blades are built around a load-bearing axis, and the airfoils are swept around it like a shell. The relative weight of the blades (kg/m^2 swept area) has halved since the early 1980s and been reduced from 3 to 1.5 kg/m^2 swept area. With carbon fibre and glass-fibre reinforced epoxy the weight has been reduced even further, down to 0.5–0.7 kg/m^2.

As turbines have grown in size, rotor blades have been made more elastic. Some of the load can then be absorbed directly by the blades instead of by the nacelle and tower. With an upwind rotor there is a limit to this elasticity as the blades should never hit the tower.

Wind turbines which are built for very cold conditions may need a de-icing system for their blades. With cold rain and fog, when the rotor is idle a crust of ice will quickly form changing the shape of the airfoil and, therefore, the aerodynamic properties of the rotor. The ice can also create imbalance on the rotor. In this case the turbine will not be able to start, the control system will not allow imbalanced running and consequently a turbine can lose a lot of valuable production during winter. To avoid this, de-icing systems can be used to melt the ice. One system uses electric heating of the blades which is governed by an ice detector on the nacelle. The anemometer and wind vane should also be heated to avoid icing in very cold conditions as they have important input into the control system of wind turbines.

Nacelle, tower and foundations

The unit mounted on top of a wind turbine tower is called the nacelle, gondola or machine cabin. Inside the nacelle is the gearbox, the generator and other mechanical and electrical components. Most large grid-connected wind turbines use conical steel towers. Smaller turbines can use a lattice tower or guyed mast. To secure turbines firmly to the ground so that they will not be blown down by strong winds, turbines are mounted on foundations of reinforced concrete. If the bedrock is solid and stable they can be bolted to the rock.

Nacelle

The nacelle of horizontal axis turbines contains a bedplate on which the components are mounted. There is a main shaft with main bearings, a generator and a yaw motor that turns the nacelle and rotor in the direction of the wind. There are also several other components depending on the model and design concept of the manufacturer.

A wind turbine of the Danish standard concept that has been used since the early 1980s (three-bladed upwind rotor and an asynchronous generator) contains a main shaft (the shaft that is turned by the rotor) and main bearings, an asynchronous generator and a gearbox that will increase the rotational speed of the rotor to the 1,010 or 1,515 rpm that the generator needs to produce electric power. There is also a yaw motor and a disc brake used for emergency stops and parking. An anemometer and a wind vane are mounted on top of the nacelle and connected to the control system of the turbine (see Figure 3.23).

The main shaft protrudes through the front of the nacelle. A rotor hub made of a steel casting is mounted at the end of the shaft. The hub is covered by a nose cone that protects the hub and reduces the turbulence in front of the rotor. Wind turbines with pitch or active stall control also have bearings for the blades and mechanical or electrical equipment to adjust the blade angles. Turbines that use a hydraulic system to rotate the blades have a hydraulic pump in the nacelle connected to a piston that passes through (inside) the main shaft out to the hub.

The purpose of the gearbox is to increase the low speed of main shaft to the speed that the generator needs: 1,010 rpm for a six-pole and 1,515 rpm for a four-pole generator. Since a large wind turbine has a rotational speed of 15–30 rpm, a significant step-up is necessary; it has to be done in several steps. Most wind turbines therefore use a three-step gearbox. Gearboxes on large turbines also need efficient lubrication and cooling, and thus also need an oil pump and oil cooling system.

Most wind turbines use so-called asynchronous generators. The generator size is specified by its nominal power, which will be attained at the nominal wind speed. When the wind speed is lower, the generator will produce less power. Many wind turbine models have two different generators, or a double-wound generator (that can alternate between four and six poles – in effect, two generators in one). The smaller generator is used for low wind speeds, and the larger for high wind speeds. There are also wind turbines that have a multi-pole synchronous generator, which can produce electric

Figure 3.23 Nacelle – Danish standard concept. The illustration shows the nacelle of a Nordex wind turbine with components of a typical pitch-regulated wind turbine with gearbox (illustration: Nordex SE)

power at low revolution speed and does not need a gearbox. The nacelle of such turbines contains very few components (see Figure 3.24).

A new concept is a so-called hybrid, which combines a multi-pole synchronous generator with low revolution speed with a robust one- or two-step planet gearbox.

Wind turbines should have two independent brake systems, one aerodynamic brake and a mechanical brake in the nacelle. The mechanical brake is mainly used as a parking brake when service and maintenance is performed in the nacelle, but it should have enough strength to stop the rotor if the aerodynamic brakes fail. A disc brake, mounted on the fast-running shaft that connects the gearbox with the generator, is used for this purpose.

Yaw control

To utilize the wind efficiently, the rotor should be perpendicular to the wind direction. When post mills were in use, the miller simply checked the wind direction, and turned the windmill towards the wind, by hand or with the help from oxen or a winch. But even in the early days of wind power, windmill engineers were very ingenious. They developed the so-called Dutch windmill with the top of the mill, the cap, disconnected from the tower so it could be turned on a slide bearing to bring the sails into the wind. A wind wheel was mounted perpendicular to the rotor. When the wind direction changed so that it hit the windmill rotor sideways, this wind wheel started to revolve. It was connected to a cogwheel that turned the cap towards the wind. When it was back in its perpendicular position towards the wind, the wind wheel stopped. This robust mechanical yaw system is still used on some smaller turbine models.

Larger turbines use yaw motors that are controlled by a wind vane. When the wind changes a specified number of degrees and this wind direction lasts a specified time, the control system will send an order to the yaw motor, which begins to turn the nacelle back to its right position in relation to the wind direction. If the nacelle has made several revolutions in the same direction, the cable from the nacelle down to the ground has to be rewound, otherwise it will be twisted off. On most models after three revolutions the turbine will be stopped by the control system, and the yaw motors turn the nacelle in the opposite direction to rewind the cables.

Towers

Most manufacturers of large turbines use conical tube towers of steel, painted white or grey, which are wider at the base than the top. In the 1980s, when the turbines were just 30 m high, they could be welded in one piece. On large turbines with hub heights of up to 140 metres, the towers are manufactured in sections, which are bolted together when they are mounted. The towers have a door at ground level, and the control system, displays and

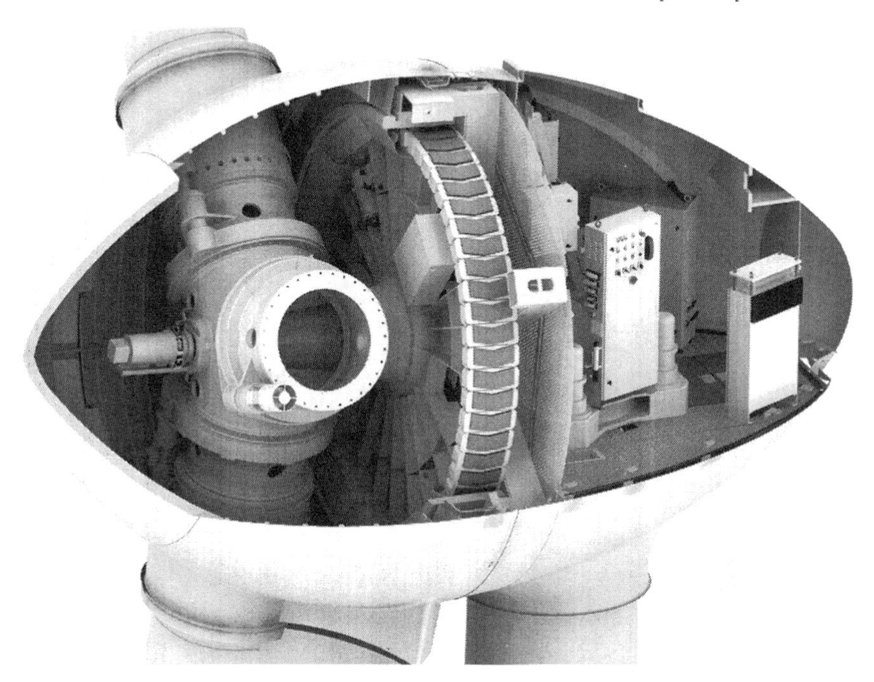

Figure 3.24 Direct drive wind turbine. Nacelle of an Enercon E-48 turbine. Wind turbines manufactured by Enercon have a very large multi-pole ring generator that is connected directly to the rotor. Inside the hub there are three electric motors for pitch control of the rotor blades, and in the nacelle there are yaw motors to align the turbine to the wind direction (source Enercon, with permission)

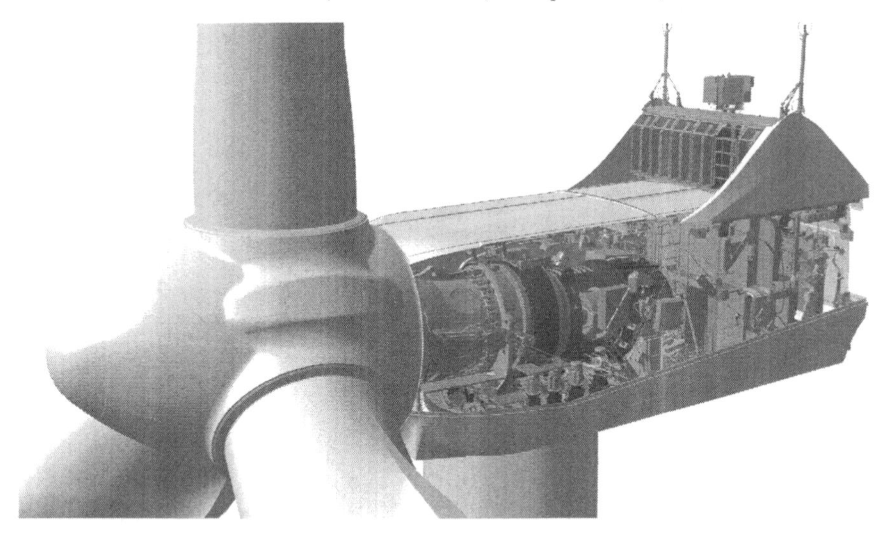

Figure 3.25 Hybrid concept. The Gamesa G128 5.0 MW turbine is a so-called hybrid. It has a robust two-step planet gearbox connected to a multi-pole permanent magnet (PM) generator (source: Gamesa, with permission)

Figure 3.26 Concrete tower. A concrete tower, like this one for a large Enercon wind turbine, is mounted in segments which are pulled together by steel wires connected to the foundation (photo: Tore Wizelius)

some electric equipment are installed inside the tower. There is also a ladder up to the nacelle on the inside of the tower.

Some manufacturers offer concrete towers as an option.

Lattice towers were often used on small wind turbine models in the 1980s. Nowadays only a few manufacturers use them, but they are still common for smaller turbines. Lattice towers have many advantages; they take less material, weigh less and cost less. Another advantage is that the wind can pass through the tower, which reduces loads on the turbine. Most manufacturers however prefer to use steel tube towers (although there are some MW turbines mounted on lattice towers), for practical and for aesthetic reasons. In Denmark, for example, it is not permitted to use lattice towers for large turbines. Small turbines often use a guyed mast as a tower.

Foundations

The foundation that the wind turbine is mounted on has two functions: to carry the weight of the turbine (and prevent it from sinking into the ground) and to act as a counter weight to prevent the turbine from falling over. The design and weight of the foundation has to be adapted not only to the size of the turbine, but also to the soil properties at the specific site where it will be installed.

On ordinary soil a 2–3 metre deep cavity is dug in the ground, forming a square or a circle 7–20 metres along the side or across the diameter. The dimensions depend on the size and weight of the turbine, hub height and ground conditions. On waterlogged ground the foundation has to be bigger to compensate for the lifting force from the groundwater.

When the bottom of the cavity has been levelled, reinforcement bars are mounted in layers separated by spacers. In the centre, a pillar is formed up to ground level which will be used as the base for mounting the tower. After that the concrete foundation is poured. The concrete then has to harden for a month before being covered by filling material and the tower mounted (see Figure 3.27).

If the wind turbine is to be installed on rock, bolts in the rock can anchor the foundation. A number of deep holes are drilled in the rock, long steel wires are inserted in the holes and expanding concrete is injected which

Figure 3.27 Gravity foundation. Forming a foundation for a 1 MW turbine. When the concrete has hardened for a month, the foundation will be covered by earth to restore the ground (photo: Bernt Johansson/Bjärke vind ekonomisk förening)

Figure 3.28 Rock foundation. Rock foundations are anchored by bolts or steel wires inserted in holes in the rock 10–15 metres deep (photo: Tore Wizelius)

fixes the wires in the rock. Each of the wires has to stand a tractive force of 30 tonnes or more, depending on the size of the turbine. A mounting base for the tower is then formed and anchored to the steel wires (see Figure 3.28).

For offshore wind turbines different types of foundations are used, depending on the character of the seabed. In shallow waters gravity foundations or monopoles are used, in deeper waters tripods. Gravity foundations are manufactured at a shipyard and are made of reinforced concrete. The seabed is levelled, the foundations are towed to the site, filled with heavy material, and submerged to the bottom.

A monopile is simply an elongation of the turbine tower. The monopile can be inserted in a hole that has been drilled in the bottom, or be driven down by a pile driver. Tripods are three-legged steel constructions.

Electrical and control systems

Most modern wind turbines are used to convert kinetic power in the wind into electrical power. The rotor transfers the kinetic power in the wind to a revolving shaft that drives a generator that generates electric power. A generator is made of a revolving part, the rotor, and a stationary part, the stator. The rotor in the generator has a magnetic field, which is created either by permanent magnets or electromagnets. When the wind turbine starts to revolve, it creates a rotating magnetic field. When this magnetic field passes the stationary coils, an electric current is induced in them and this current can be fed into the power grid.

Most generators generate alternating current, AC. This means that current and voltage will change directions several times during each revolution of the rotor. The frequency of the AC current, the number of periods (from neutral to positive and then through neutral to negative and back to neutral in a sine curve), depends on the rotational speed of the generator. To get a constant frequency from the generator, the rotational speed of the wind turbine rotor has to be fixed. If the rotational speed varies, the frequency and voltage will vary as well.

Electric systems on wind turbines

Wind turbines with generators that are directly connected to the power grid have a rotational speed that corresponds to the grid frequency. In Europe the frequency is 50 Hz, 50 cycles per second (in the US the frequency is 60 Hz). A simple generator with only two poles (N and S) would then need a rotational speed of 50 × 60 seconds = 3,000 rpm (revolutions/minute) to give 50 Hz. The number of poles in a generator can however be increased, so that there is more than one cycle during one revolution. A generator with four poles gives 50 Hz with 1500 rpm. (The relation between rotational speed and number of poles is calculated as $n = 6{,}000/p$ where p is the number of poles.) Most mass-produced generators have four or six poles.

In the power grid a three-phase alternating current is used, which means that three alternating currents run parallel with each other. These three currents are displaced a third of a period from each other. The generator has to generate three separate AC currents, and needs a set of poles for each of them.

Some manufacturers use ring generators. These have a large number of poles, not 4 or 6 but 64 or 96 or other number of poles, depending on the size and usage of the generator. They also have a large diameter and can therefore be run with a low rotational speed (the peripheral speed increases with the diameter, just like the tip speed of the wind turbine rotor). By increasing the number of poles and the diameter, the rotational speed necessary to generate electric power in a reasonably efficient way can be reduced to the speed of the turbine rotor. With this design concept, a gearbox will not be needed.

Two basically different kinds of generators are used in wind turbines; synchronous and asynchronous. A synchronous generator can be connected to the grid, or work without grid connection (connected to a local grid, battery storage, and local loads such an electric water pump). An asynchronous generator has to be connected to a grid to function, since it is dependent on it to magnetize the rotor and it is governed by the grid frequency. When there is a perfect match between the rotational speed of the generator and the grid frequency, the generator runs idle. To generate power, its rotational speed has to be asynchronous; a little higher than the frequency. An asynchronous generator with a nominal rpm of 1,000 rpm, has to be run at 1,010 rpm to produce full power (see Figure 3.29).

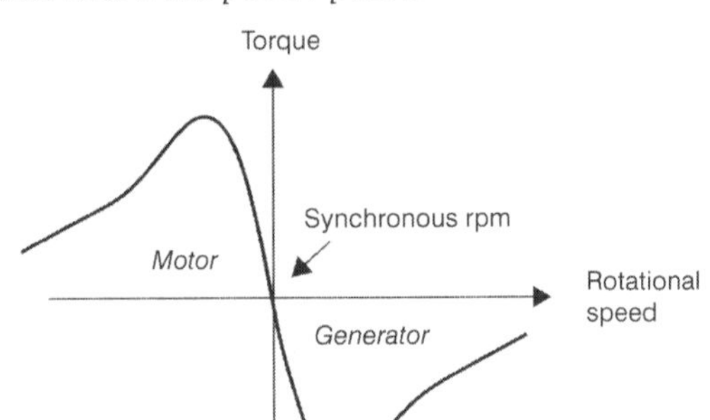

Figure 3.29 The moment curve of an asynchronous machine. The diagram shows the relation between the torque of the shaft and the rotational speed for an asynchronous machine

Asynchronous generators are installed in turbines with a fixed rotor speed (rpm). Actually, the rotational speed is not completely fixed. If the rotational speed is lower than the synchronous speed, the machine will give a positive torque; it works as an electric motor. If the shaft is revolved (by the wind turbine rotor) a little faster than the synchronous rpm, it will start to function as a generator and feed power to the grid. The normal operating range from the synchronous rpm up to the nominal rpm, is about 1 per cent higher (at torque of 100% in Figure 3.29). When the wind speed increases and the torque increases, the rotor tries to accelerate, but instead the force of the magnetic field in the rotor is increased, which results in an increased power in the generator and more power fed to the grid.

The operating rpm deviation from the synchronous rpm is called *slip* and is expressed as percent deviation from nominal rpm. This slip will be highest at the nominal power of the generator, and is then around 1 per cent. When the generator works at full capacity, the torque on the rotor shaft cannot be allowed to increase. If the wind speed gets stronger, some of the power that turns the rotor must be diverted (or spilled) by some means of power control – pitch or stall – otherwise the generator will be overloaded. If a situation of too high power generation lasts too long, the generator will break down (the windings will melt from the heat, for example).

For shorter periods the nominal power can be exceeded. This happens, for example, before the pitch control system has adjusted the blade angle to a new higher wind speed. For short periods the generator can also be run as a motor to avoid the turbine cutting in and out too often when the winds are close to the cut in wind speed.

A *synchronous* generator does not need a grid connection to be able to produce power. It can be used for a local grid, and be directly connected to an electric water heater, where electric resistors that create the heat can be connected stepwise and adapt the rotor rpm to the wind speed by changing the load. By doing this the turbine can be kept at an efficient tip speed ratio. The frequency will vary, but for a heating load this does not matter.

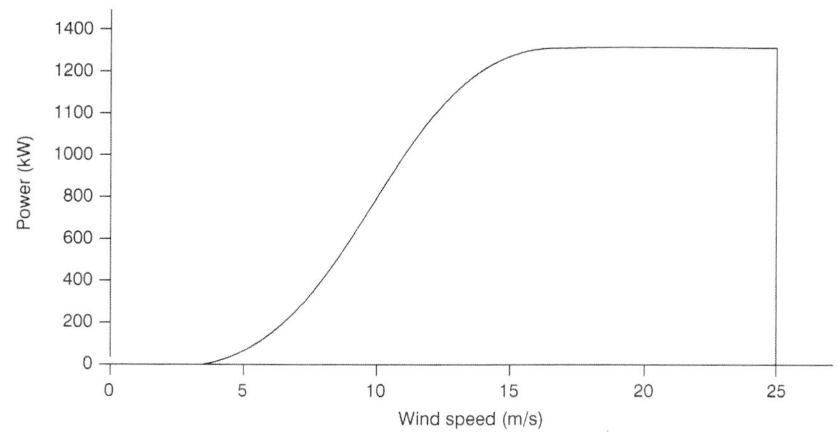

Figure 3.30 Power curve for a wind turbine. The power curve shows how much electrical power a wind turbine will produce at different wind speeds. This turbine starts to produce power at 4–5 m/s and reaches the nominal power, 1300 kW, at 16 m/s. With higher wind speeds the power levels out, and at 25 m/s, the cut-out wind speed, the turbine is stopped by the control system

Relation between wind speed and power

The size of generator is defined by its nominal power. On a wind turbine this power will be reached at the nominal wind speed (12–16 m/s, depending on the manufacturer, model, site etc.) and higher wind speeds. At lower wind speeds the power is significantly lower. The relation between wind speed and power is shown in a power curve (see Figure 3.30).

At a certain maximum wind speed (in most cases 25 m/s) the wind turbine will be stopped and disconnected from the grid. The loads on the turbines will be reduced when the turbine is stopped and parked, and the rotor does not revolve. In this parked state, wind turbines will survive winds up to hurricane force (60 m/s). The maximum wind speed that turbines are designed to stand is called the *survival wind speed.* The lost production from periods with wind speeds above the cut-out wind speed, more than 25 m/s, are negligible, as these wind speeds are very rare and occur only for short periods of time.

The American Charles F. Brush built the first wind turbine that produced electric power in 1887–88. It operated for 20 years and was used for battery charging. Wind turbines for electric power production were also developed in Denmark around the turn of the twentieth century. These turbines were made for stand-alone systems and were used to charge batteries, heat water and run electric water pumps. The early turbines had DC generators. In this kind of generator the power is produced in the rotor, and it is transmitted from the generator by a revolving contact – a commutator. A serious drawback with DC generators is that the commutator wears out, increasing the need for servicing, and this kind of generator is therefore seldom used nowadays.

If an AC generator is used to charge batteries, the current first has to be converted to DC by a rectifier. There is a considerable niche market today for this kind of wind turbine with a power from 100 W to a few kW. They are used as battery chargers on caravans, sailing yachts and holiday cottages, and for lighthouses, telecommunication masts and offshore oil platforms and in areas with no grid connection. The rotational speed on small turbines is so high that no gearbox is needed. Most of them have direct drive synchronous generators with permanent magnets.

Asynchronous generators with fixed speed

The first commercial wind turbines for grid connection, which were installed at the end of the 1970s and the beginning of the 1980s, used standard asynchronous generators that were available on the market, and rotors with stall control. They had a fixed speed and used a very simple electrical system (see Figure 3.31).

This simple design caused some problems for grid operators, who claimed that wind turbines had a negative impact on the power quality, especially when the number of turbines increased. The impact when a wind turbine generator is cut in to the grid, is actually the same as when an electric motor of the same size is started. When the turbines start and are connected to the grid, the generator needs reactive power to magnetize its rotor; and when the generator is connected to the grid, a short but powerful current, a spike, sets in and the voltage drops for a split second. During operation the turbines produce active power, but also consume some reactive power from the grid. This will put unnecessary demand on the grid capacity.

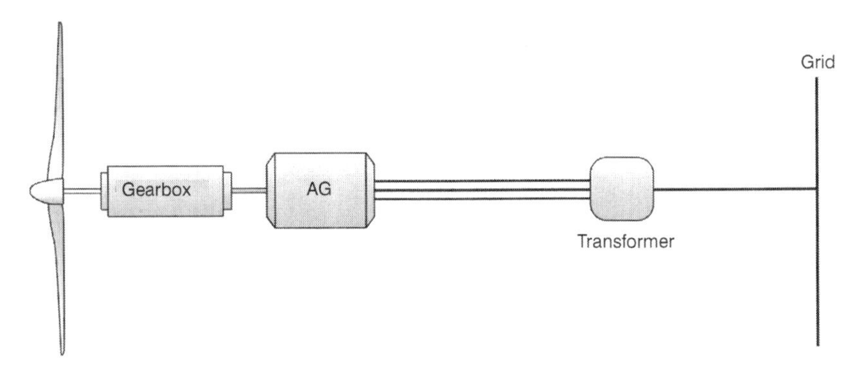

Figure 3.31 Asynchronous generators with fixed speed. The load in an asynchronous generator (AG) will increase in proportion to the power transmitted from the rotor, so that the speed will remain fixed (with very minor variations, slip). To increase the rotational speed to 1,500 rpm a gearbox is necessary (illustration: Typoform)

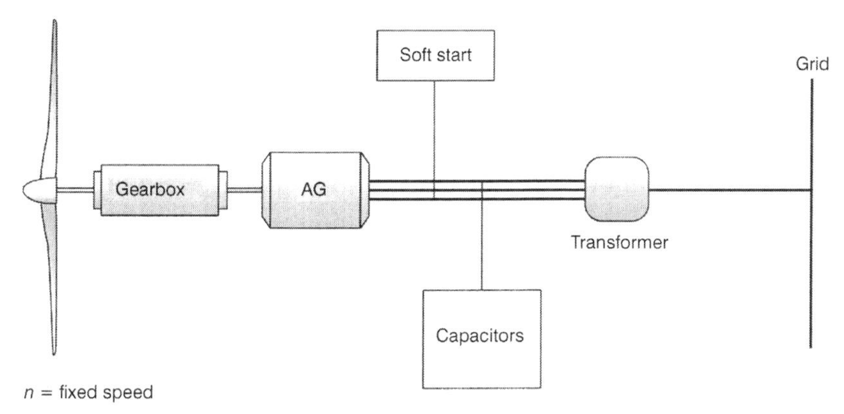

Figure 3.32 Grid-adapted asynchronous generator. To eliminate spikes when the turbine is cut in to the grid, soft start equipment is used; and to reduce the demand for reactive power, capacitors are installed

Soft start generator

At that time the turbines were quite small, and the disturbances on the grid were within tolerable limits. However, additional equipment that would reduce these problems was soon installed on the turbines: *capacitors* to reduce the reactive power consumption; and so-called *soft start equipment*, a couple of thyristors to reduce the cut-in currents (see Figure 3.32).

When a generator runs on partial load (for example a 500 kW generator produces just 100 kW), the efficiency of the generator will be considerably lower than when it runs on full power. With a fixed rotational speed that has been set to get a good tip speed ratio in relatively strong winds, the efficiency of the rotor will be comparatively low at low wind speeds, since the tip speed ratio will be too high. At the same time the generator that runs on partial load will have lower efficiency.

Double generators

To utilize the power in the wind in a more efficient way, especially in low wind areas (with low average wind speed), some manufacturers began to install two separate generators, or a double wound generator (that works as two generators in one). A small generator with six poles and 1,000 rpm runs on full load in low wind speeds with a lower rotational speed (better tip speed ratio) on the rotor. When the wind speed increases over a certain limit (~ 7 m/s) the large generator, with four poles and 1,500 rpm, takes over and the rotational speed increases to a higher fixed level (see Figure 3.33). By reducing the rotational speed at low wind speeds, the aerodynamic swish noise from the rotor will be reduced as well, which is an important advantage since the noise generated is the most important limiting factor for siting of turbines in inhabited areas.

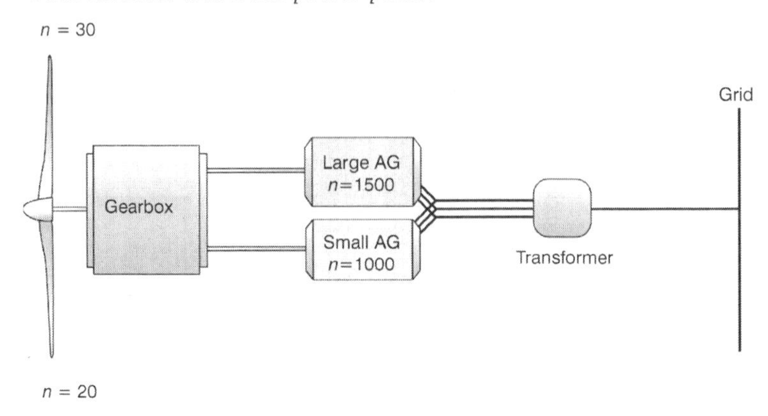

Figure 3.33 Wind turbines with two generators. To utilize low wind speeds more efficiently, turbines with two generators were developed: one small generator for low winds (and a low rotational speed on the rotor); and one large generator with the nominal power of the turbine for higher wind speeds (and higher rotational speed on the rotor) (illustration: Typoform)

Variable speed

To get the maximum efficiency in utilizing the power in the wind, the rotational speed of the wind turbine rotor has to be proportional to the wind speed. With a variable speed the tip speed ratio can be kept at the optimal level for all wind speeds. Turbines with a high tip speed ratio have a wide range for the tip speed value; it can vary quite a lot around the optimal value and still have good efficiency. The introduction of two generators was the first step towards variable speed turbines.

If a generator is run on variable speed the frequency of the electric power will also vary. The electric current has to be adapted to fit the grid. To solve this problem, the generator is 'disconnected' from the grid. The AC from the generator is first rectified to DC, and then converted to AC by an inverter which will give the current the same frequency and voltage properties as the grid. Wind turbines with variable speed have been available on the market since the late 1980s. They use a synchronous generator combined with power electronics: a frequency converter (see Figure 3.34).

Direct drive generator

Multi-pole synchronous generators, so-called *ring generators*, have been used for a long time in hydro power stations. The advantage of this type of generator is that it can be operated with a low rotational speed. In a wind turbine a ring generator can be driven directly by the rotor without an intermediate gearbox. The wind turbine can have a *direct drive* generator.

The German manufacturer Enercon first introduced this design in 1992, with the model E-40. This ring generator has a large diameter, around

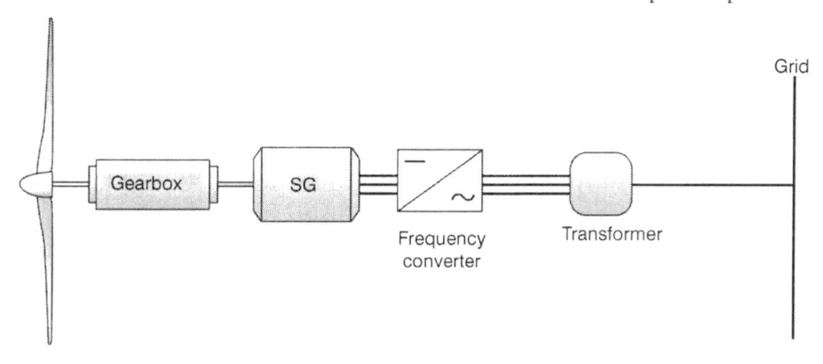

Figure 3.34 Variable speed turbines. Wind turbines with variable speed can use a synchronous generator (SG). The AC current is rectified and then inverted back to AC with the same frequency as the power grid by a frequency converter (illustration: Typoform)

four metres, and about 60 poles instead of four or six that is normal on a standard generator. The ring generator can produce electric power with the same rotational speed as the rotor. Enercon turbines have variable speed and use power electronics to adapt the electric power to the grid (see Figure 3.35).

One important advantage of a direct drive generator is that there is no gearbox that needs maintenance. The main disadvantage is that large ring generators are very heavy, so the weight of the turbine increases. Some manufacturers have developed a hybrid; a wind turbine with a smaller multi-pole ring generator combined with a simple and very robust one or two step planet gearbox. Nowadays permanent magnets (PM) are used in some multi-pole generators.

With variable speed the turbine rotor can utilize the wind more efficiently. There are also other advantages of variable speed. The wind is always more

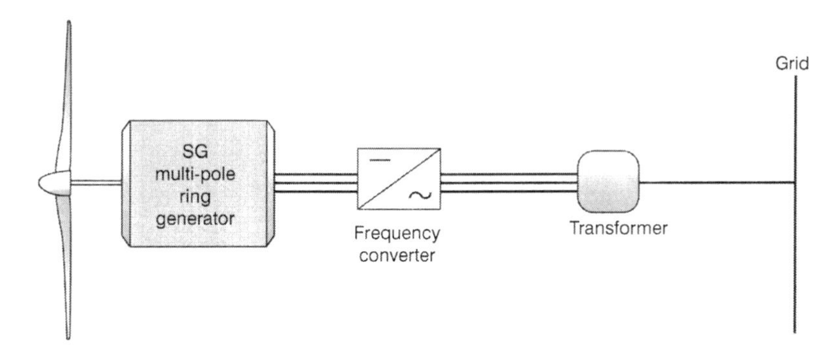

Figure 3.35 Turbine with direct drive ring generator. A multi-pole ring generator can be driven directly by the rotor since it has a low rpm. Because the synchronous generator operates at a variable speed the current has to be converted to the grid voltage and frequency by power electronic equipment (illustration: Typoform)

or less turbulent and the loads on the turbines will therefore change all the time, which causes great strain on all components. If the rotor speed is allowed to vary, the power from gusts can be absorbed by the rotor, by increasing the rotational speed (it accelerates), and not be passed on to the main shaft and other components. When turbines get bigger, it will be more important to reduce the loads on the turbines.

Generator with slip

Manufacturers who use asynchronous generators have designed their turbines so that the rotational speed is allowed to vary within certain limits. The technical solution to do this is to change the resistance in the rotor windings, or to control the currents in the rotor windings by power electronics. Vestas developed a system named OptiSlip. This has made it possible to increase the 'slip' of the generator rpm from 1 to 10 per cent. When wind gusts give sudden increases of power, the turbine rotor can increase its rotational speed by 10 per cent, without affecting the frequency or power output of the generator. The excess power is turned into heat.

The next step in the technical development of the asynchronous generator was to use a slip ring with a rotor cascade coupling (see Figure 3.36). Vestas, Nordex and several other manufacturers use this system. In an asynchronous generator currents are induced also in the generator's rotor. This current has a low frequency, which is governed by the slip of the generator's rotational speed. This current can be transmitted through slip rings and the frequency can be converted to the grid frequency. This concept makes it possible to govern the rotational speed over a wide range; it can be variable. In a Nordex 2.5 MW turbine the rpm can vary from 10–18 rpm. With this solution the frequency converter can be much smaller, since only a small

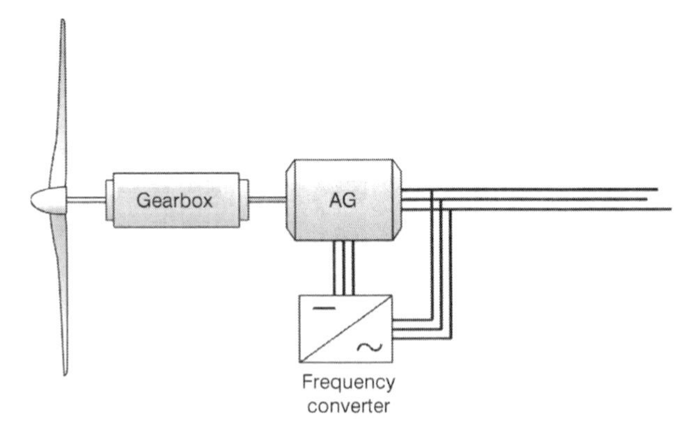

Figure 3.36 Asynchronous generator with rotor cascade coupling and frequency converter. Rotor cascade enables variable speed; most of the power is generated in the stator windings. Only a small proportion from the rotor windings has to be converted to the grid frequency (illustration: Typoform)

proportion, about 20 per cent of the nominal power of the turbine, needs to be converted. Another advantage with this system is that the reactive power can be controlled.

Grid connection

The voltage level of large wind turbines is in most cases 690 V, so-called industrial voltage. They can be connected to a factory without a transformer. Smaller turbines, up to 300 kW that were common at the beginning of the 1990s, have a voltage of 400 V and can be connected directly to a feeder cable to a farm or a house. Usually wind turbines are however connected to the power grid, through a transformer that increases the voltage level from 400 or 690 V to high voltage, which normally is 10 or 20 kV in the distribution grid. Large wind power plants also have an internal grid, and a substation connected to the regional grid with higher voltage.

For small and medium-sized wind turbines a suitable transformer is installed on the ground next to the tower. In large wind turbines the transformer is an integrated component of the turbine itself. It can be installed in the 'cellar' of the tower, below the level of the door. Some manufacturers have installed the transformer in the nacelle, where it also acts as a counterweight to the rotor. The cables that transmit the power from the nacelle to the ground, through the tower, can be thinner, since high voltage cables need much less cable area than low voltage cables.

Electric systems in wind turbines have seen much development since the early 1980s and they have been adapted to be 'kinder' to the grid. In modern wind turbines the properties of the electric power that is fed into the grid can be governed by power electronics, so that they will have the phase angle and reactive power that the grid needs at the point where the wind power plant is connected. Wind turbines, which caused problems of power quality for grid operators, can now be utilized to improve the power quality in the grid.

But, the power electronics used to solve the old problems have created some new ones (this is the case also with the power electronic equipment used in factories and households). The electronic equipment generates harmonics, currents with frequencies other than 50 Hz. This has a negative impact on power quality. This 'dirt' can, to some extent, be 'cleaned' by different kinds of filters, but this type of equipment is expensive, and it seldom manages to take care of all the 'dirt'.

The control system of wind turbines

Every little movement the wind turbine makes is governed by modern computer technology, and 'spikes' and other 'pollution' in the current are screened out by power electronic equipment. Turbines are connected through a modem to the owner's and the manufacturer's offices, which get operational data displayed directly on their computers. If a malfunction happens, an operational alarm will alert the owner or operator of the turbines.

The control system fulfils three different functions: operation control, monitoring and operations follow-up (see Box 3.4).

A computer, which in most cases is installed on the ground level inside the tower, governs the control system of a wind turbine. Often there is a terminal with a display also in the nacelle which is used during maintenance. Data collected from the anemometer, machine components and from the grid are transmitted by fibre-optic cables.

The wind is always varying – the wind direction as well as the wind speed will change almost every second. To maintain efficient production,

Box 3.4 Control system functions

The control system of a wind turbine has three different functions:

Operation control

The computer in the wind turbine collects data from the wind vane and anemometer, and if the nacelle has to yaw into wind, it sends a signal to the yaw motor to start working. On turbines with pitchable blades the control system manages the blade angle adjustments. It also regulates when the generator is connected to or disconnected from the grid.

Monitoring

Sensors will check the temperature in the gearbox, generator and many other components, vibrations in the rotor and nacelle, grid voltage and many other parameters. When a value exceeds its tolerance level, the turbine will be stopped, and an alarm signal is sent to the owner/operator by telephone or staff locator. If the fault isn't serious, the operator can restart the turbine remotely by computer.

Operations follow up

The computer collects data about production, wind speed, outages and other parameters. The information is processed and presented in visual form, graphs, tables etc. on the display in the turbine, or on remote computer.

All medium and large wind turbines on the market have advanced computerized control systems.

the turbine rotor should be perpendicular to the wind direction. The control system continuously checks the wind speed and wind direction. The information is processed by the computer, which can order the yaw motor to turn the nacelle a specific number of degrees. On turbines with pitchable blades, the control system calculates when and how much the blades shall be adjusted.

The wind turbine cannot follow all the unpredictable changes of the wind; if it did the nacelle would constantly move to and fro, and the yaw motor would soon be worn out. Reliability in operation is a very important property for wind turbines, as they are often sited in remote places. They should be in operation day and night, all year around, without expensive servicing and maintenance. The control system has to be designed not only to optimize production but also the useful life of the turbine and to protect the turbine against damage from power outages or component break-down or malfunctioning.

Therefore the control system will not order the yaw motor to turn the nacelle until the change of wind direction seems to be lasting. The new wind direction has to change at least a preset number of degrees, and keep that direction for a preset number of seconds or minutes, before the order is given to the yaw motor to adjust the nacelle. The control program also has to keep count of how many revolutions the nacelle has made. After three full revolutions the cables that hang down through the tower will have been twisted into a tight spiral. The control system then stops the turbine and gives the yaw motor order to rewind the nacelle and the cables, before the turbine is restarted again.

The control system also keeps a close watch on all the functions of the turbine. The control system is like the brain with a nervous system of fibre-optic cables with sensors that check the temperature in the gearbox and generator, the pressure in the hydraulic system, vibrations in machine components and rotor blades, the voltage and frequency in the generator and the grid, and many other parameters. These data are saved in the computer's memory for a couple of days, to enable a thorough analysis in the event of operational disruption.

Wind turbines often have to use the emergency brake, even if there is no fault in the turbine itself. A power failure in the grid is the most common cause for this. Another trivial reason for stoppages is that a sensor used for monitoring breaks down. If a serious fault has occurred, the turbine has to be restarted at the site. The operator has, however, received a fault report on his computer and knows which spare parts he needs to repair the turbine. Often this is some component of the advanced control system that breaks – a printed circuit board or a sensor. The control system itself is one of the most vulnerable parts of modern wind turbines.

Data technology has developed almost as fast as wind power technology, and the cost of advanced computer hardware and software has decreased. By using advanced software to control turbines, efficiency can be increased and this software is one of the manufacturers' most valuable assets. Some manufacturers have connections via modem to wind turbines installed in different locations, and can upgrade the software that controls the operation of the turbines remotely.

The demand for reliability and technical availability has increased even more when wind power has been developed offshore. The next step will be to install advanced sensors that can give information about the status of components, and give warnings when they start to be worn out, so that they can be replaced or repaired before they break. Duplicate sets of sensors, so that a faulty sensor does not cause a halt in the turbine, have already been introduced on many models.

In many countries information about the production, fault reports, etc. are collected and published in reports, or on the internet. Making these data publicly available has been very valuable for the development of wind power.

Efficiency and performance

How much energy a wind turbine can produce depends on a number of factors – the area swept by the rotor, the hub height and how efficiently the turbine can convert the kinetic power of the wind into electric power. Equally important, of course, is the mean wind speed and the frequency distribution at the site where the wind turbines are installed.

Ever-larger rotors

The power of the wind is $P = \frac{1}{2} r A v^3$. The power is proportional to the rotor swept area A and the cube of the wind speed v. The rotor swept area of wind turbines has increased at a steady rate and, as a consequence, has the rated power of the turbines. From the early 1980s the power of wind turbines has doubled every 4–5 years on average (see Figure 1.3).

The rotor areas of commercial wind turbines have increased from scarcely 200 m^2 in 1980 to 5,000 m^2 at the beginning of this century and to 13,000 m^2 in 2014. The nominal power increased from 50 kW to 6 MW in the same period. Annual production has increased at a similar growth rate, from about 90 MWh/year to 15,000 MWh/year. The size of the turbines (nominal power of the turbines dominant in the market) has doubled about every 4–5 years. In 2015 there are wind turbines with a nominal power of 5 MW available, and if the development continues in the same way there will be 8–10 MW turbines within a few years.

The rotor swept area has not increased at the same rate as the nominal power of the turbines. This is because towers also have increased in height and a larger rotor needs a higher tower. Since wind speed increases with height, the turbines can capture more power and so it makes sense to use generators with higher ratings.

It is not easy to increase the size of a wind turbine and keep the costs on a competitive level. When the size of wind turbines increases, the larger turbines have to produce power at a lower price; otherwise it would be pointless to increase the size. The problem is that when the radius of the rotor is increased, the swept area will increase by the square of the radius, but the volume and weight will increase by the cube. The rotor blades and all other components have to be scaled up – length, width and thickness. If a small turbine is scaled up to a larger size just by increasing the proportions of all components, the weight will obviously increase much faster that the swept area. And the price of material is proportional to the weight of the wind turbine.

This seemingly impossible relationship has however been overcome by advanced engineering, more accurate calculations of loads, development of new design concepts, control strategies and materials. So to grow, turbines have had to go on a diet to reduce the unavoidable increase of the weight.

Two factors have made this possible. The cost efficiency has been increased by the increase of power with height. The second factor is that the early turbines of the 1980s were over-sized, which has made it possible to cut some weight from most components. In some cases, however, the diet was too strict, which caused some very comprehensive and expensive (for manufacturers as well as owners) retrofits of gearboxes in some wind turbine models (see Figure 3.37).

It is impossible for a wind turbine to utilize all the power in the wind. How large a proportion of the total power in the wind a turbine will utilize is indicated by the power coefficient C_p. The maximum value of C_p is 0.59

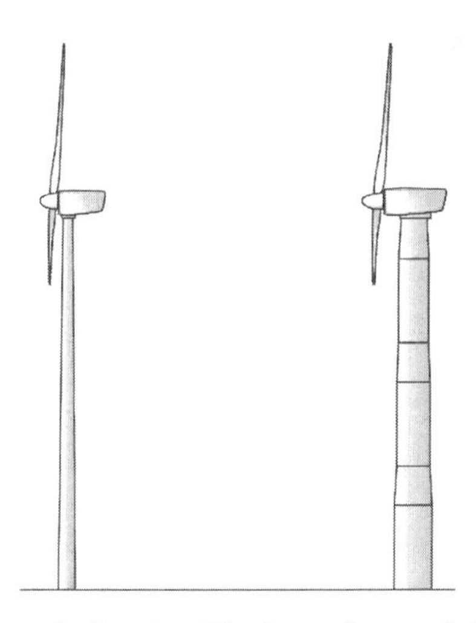

Figure 3.37 Growth through slimming. The figure shows a 2 MW turbine to the left compared to a 55 kW turbine from 1980 that has been up-scaled to a comparable size (2 MW). The 2 MW turbine is much slimmer and more compact (source: Stiesdal, 2000)

according to Betz' law. The power coefficient varies with the wind speed, and on most turbines the maximum value that can be attained is 0.45–0.50 at a wind speed of 8–10 m/s. Most turbines are optimized for these wind speeds, which often are the most frequent wind speeds in the frequency distribution for a year. The tip speed ratio on most turbines is set below the optimum value, to limit the swishing noise from the rotor blades.

To convert the power from the revolving rotor to electric power, it will pass through a gearbox and a generator; or for direct drive turbines, through a generator and an inverter. In this conversion some power will be lost as heat. Also the efficiency of the gearbox, generator and power electronic equipment will vary with the wind speed.

Generator efficiency

A generator is most efficient when it runs on its nominal power. On a wind turbine, the generator most of the time runs on lower power, as the wind speed is lower that the nominal wind speed. The generator then runs on *partial load.* On a standard generator the efficiency will then be reduced (see Box 3.5).

The gearbox

On a large modern wind turbine, the rotor has a rotational speed of 6–30 rpm, while the generator will need to rotate at 1,515 rpm. To increase the

Box 3.5 Generator efficiency

Reductions of generator efficiency at partial load

Per cent of full load	*5*	*10*	*20*	*50*	*100*
Efficiency	0.4	0.8	0.90	0.97	1.00

There is also a relation between the physical size of a generator and efficiency; the efficiency increases with the size of the generator, since the heat losses are reduced.

Relation between size and efficiency

Nominal power kW	*5*	*50*	*500*	*1,000*
Efficiency	0.84	0.89	0.94	0.95

A 1 MW turbine that runs on 20 per cent of its nominal power (200 kW) has an efficiency of 0.95 × 0.90 = 85 per cent. The relations between efficiency, size and partial load can vary between different manufacturers and models. The figures in the tables are typical examples.

speed a gearbox is used. If the turbine rotor runs at 30 rpm a gear change of 30/1520 = 1:50.7 will be needed. One revolution of the main shaft has to be increased to 50.7 revolutions on the secondary shaft that is connected to the generator. The gearbox has one fixed gear change ratio (unlike gears in a car). With double generators with 1,000 and 1,500 rpm respectively, different speeds from the turbine rotor are used for the two generators. In this case the rotational speed for the small generator for low wind speeds (for the small generator with six poles) will be 20 rpm. A hybrid using a one- or two-step gearbox and a multi-pole synchronous generator does not have to increase the rpm as much, which is one reason why this concept is used.

A gearbox generally has several steps, so that the rotational speed is increased stepwise. The losses can be estimated to 1 per cent per step. In wind turbines three-step gearboxes are usually used and the efficiency of the gearbox will then be around 97 per cent.

Wind turbines with a direct-drive generator and variable speed don't need any gearbox. Instead the frequency and voltage of the electric current will vary with the rotational speed. The current therefore has to be rectified to DC (direct current) and then converted by an inverter to alternating current (AC) with the same frequency (50 Hz in Europe, 60 Hz in the US) and voltage as the grid. The efficiency of such an inverter is about 97 per cent; the losses will therefore be about the same as for a turbine with a gearbox.

Overall efficiency

The overall efficiency for a wind turbine is the product of the turbine rotor's power coefficient C_p and the efficiency of the gearbox (or inverter) and generator.

$$\mu_{tot} = C_p \cdot \mu_{gear} \cdot \mu_{generator}$$

Sometimes C_p is set to 0.59 and μ_{rotor} (μ_r) is used to show how large a share of the theoretically available power the rotor can utilize. If the power coefficient C_p = 0.49 the rotor efficiency will be μ_r = 0.49/0.59 = 0.83.

The efficiency of a wind turbine varies with the wind speed. When the wind speed is below the nominal wind speed, the efficiency of the generator will decrease, and if the turbine has a fixed rotational speed, the tip speed ratio will change so that the C_p is also reduced. When the wind speed is higher than the nominal wind speed, some of the power in the wind will be spilled, so that an ever-smaller share of the power in the wind will be utilized and C_p will decrease progressively. Wind turbines are used to convert wind to electric power, and therefore another coefficient is also used, C_e, which shows how large a proportion of the power in the wind is converted to electric power at different wind speeds (see Figure 3.38).

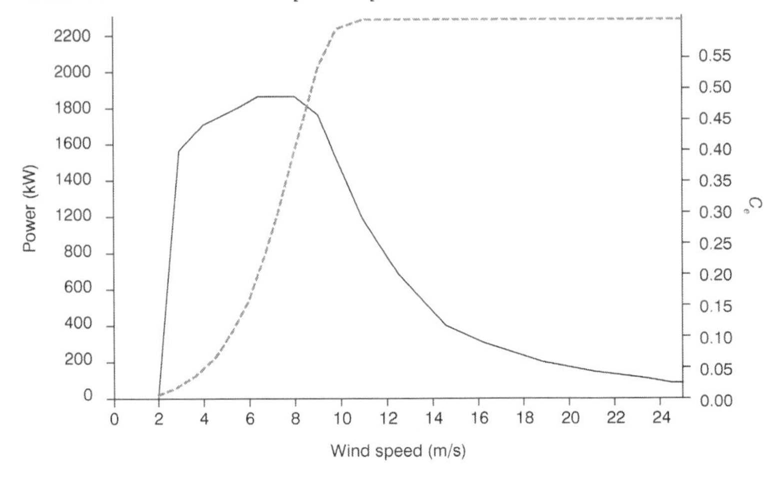

Figure 3.38 C_e graph for Siemens SWT 2.3-113. The C_e diagram is the solid line with values on the right vertical axis. It shows how large a proportion of the power in the wind is converted to electric power at different wind speeds. The turbine is most efficient at wind speeds between 6 and 8 m/s (source: EMD, 2014)

Power curve

A power curve shows how much electrical power a wind turbine will produce at different wind speeds. This curve can be calculated if the efficiency of the different components at different wind speeds is known. This curve has however also to be verified by measurements when the turbine is in use. There are very specific rules for how such measurements should performed and independent certification institutes or companies make these measurements to verify the power curve.

The wind speed is measured by an anemometer at hub height on a measurement mast erected at a suitable distance from the turbine, and the power from the turbine is measured simultaneously. During the measurement period all wind speeds have to occur for a specified time, from calm to more than 25 m/s. The results from these measurements are made into a diagram with the wind speed on the x-axis and the power on the y-axis. Each measurement results in a dot, and together they form something that is far from an even curve, it rather looks like a swarm of mosquitoes.

The reason for this is that there is a short delay before the rotor can catch a gust and turn it into an increase of power, and when the wind slows down the power is kept on the present level a short time due to the force of inertia of the revolving rotor; if the wind suddenly calms completely the rotor will continue to revolve a turn or so before it stops. By calculating the average power for different wind speeds, a smooth, even curve will be formed (see Figures 3.39–3.42).

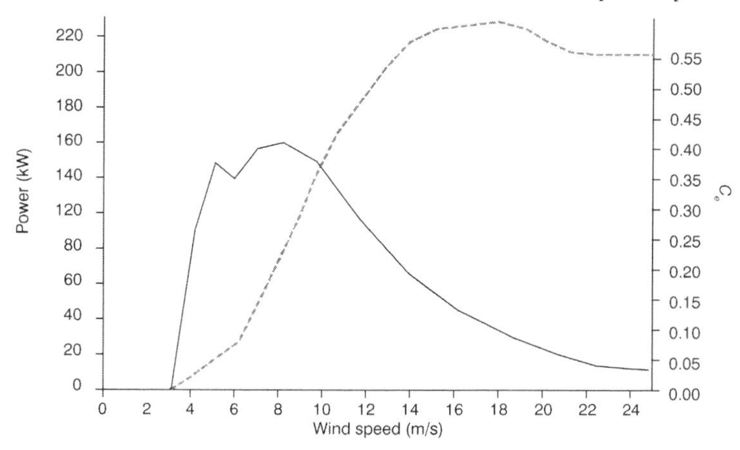

Figure 3.39 Power curve for NEPC 225-40. This 225 kW turbine from the Indian manufacturer NEPC is stall controlled. The power increases rather smoothly up to 80 kW at 8 m/s, where the blades begin to stall (this is seen on the C_e curve). The power continues to increase up to 225 kW at 16 m/s, but decreases after that when the stall increases. With fixed rotor blades, only the form of the airfoil controls the power, and it is hard to get a level curve above the nominal wind speed. On earlier models the power dropped more dramatically, but by developing the airfoils it has been possible to make the curve straighter (source: EMD, 2014)

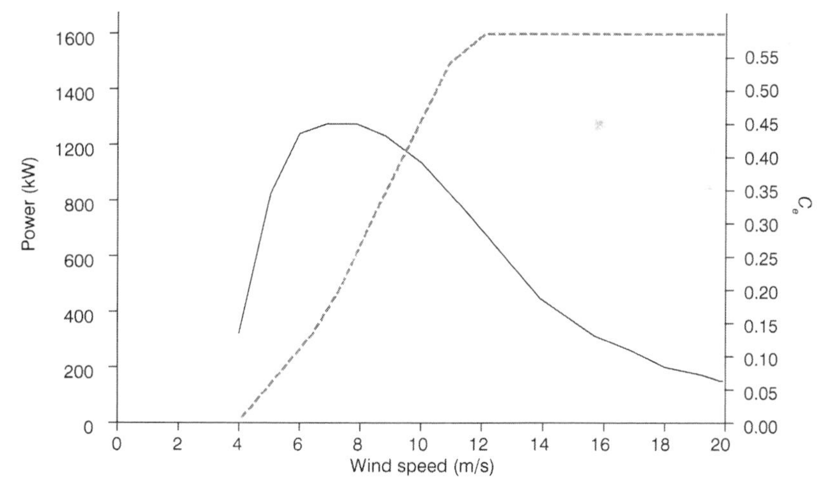

Figure 3.40 Vestas V82 1650 – active stall control. Vestas V82 controls the power by active stall, the rotor blades can be adjusted to control the stall so that the power curve becomes straight above nominal wind speed (source: EMD, 2014)

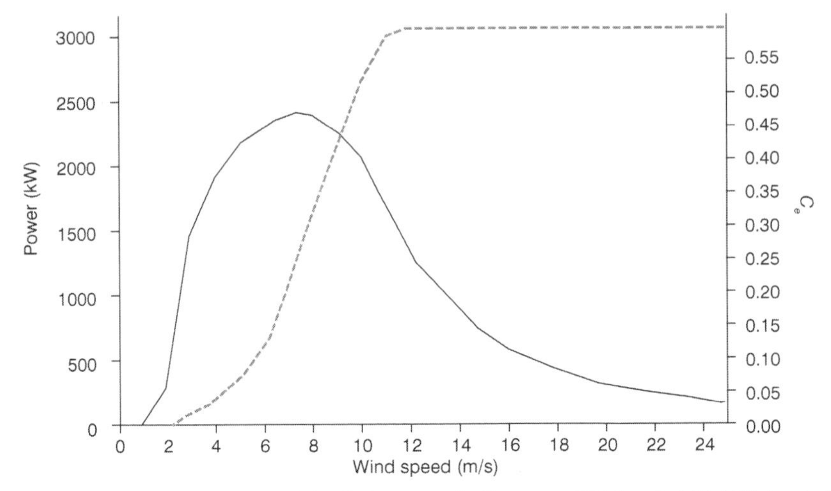

Figure 3.41 Enercon E115 with pitch control. Enercon E115 – 3,000 kW is pitch controlled and has a smooth power curve that is straight above the nominal wind speed, up to 25 m/s where it is cut out from grid and stopped (source: EMD, 2014)

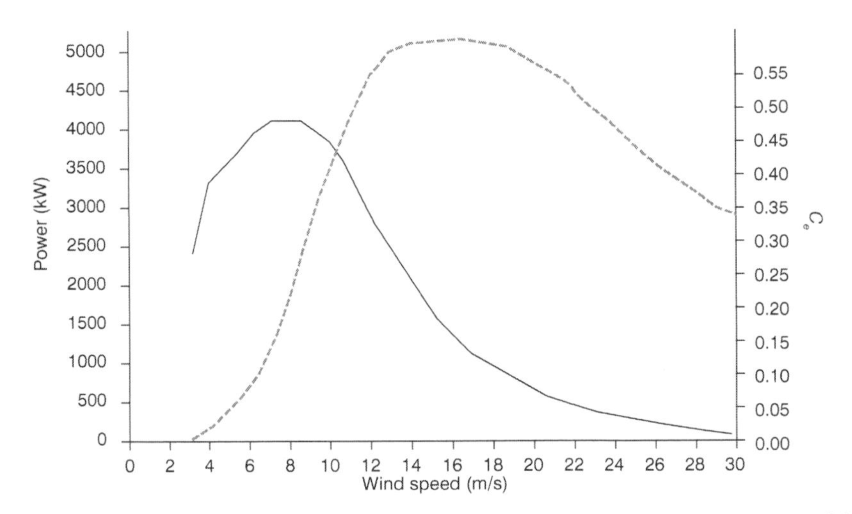

Figure 3.42 Gamesa G128, 5 MW, with pitch control. This very large and powerful wind turbine with 128 metre rotor diameter reduces the power when wind speeds increase above 18 m/s to reduce loads on the turbine. It continues to produce power up to 30 m/s

Production capacity

To calculate how much electricity a certain wind turbine can be expected to produce at a specific site, it is necessary to know the frequency distribution of the wind speed at that site, at the hub height of the turbine. A histogram or table of the frequency distribution shows how many hours a year different wind speeds occur. The power curve for the turbine shows the power it will give at these different wind speeds. The estimated production in a year can then be calculated by multiplying the frequency distribution by the power curve.

How much a wind turbine will produce depends to a large degree on the wind conditions at the site where the turbine is installed. This makes it quite difficult to compare the efficiency of different turbine models, since they are installed at different sites. It is not only the mean wind speed at the site that matters, the frequency distribution is equally important. Finally it is not the *technical efficiency* that is decisive, but the *cost efficiency*.

It is, however, possible to tailor a turbine for a specific site. There are several options for hub heights. Close to the coast or offshore, a low tower may suffice; on an inland site the tower for the same turbine has to be much higher to get the same production. For a site with a low mean wind speed, it pays to choose a turbine with a large rotor and a low nominal wind speed. For turbines in offshore wind farms the power of the generator, and the nominal wind speed, can be increased instead. Besides the hub height, turbines can be adapted to a site by choosing a good relationship between the size of the rotor and the size of the generator.

Key figures for efficiency

Different key figures are used to estimate efficiency in wind power statistics.

- Annual power production/nominal power expressed in kWh/kW
- Annual power production/swept area expressed in kWh/m^2.

In both cases mean values for a year are used. However, none of these key figures gives a good estimate on the efficiency of a turbine. A turbine with a large rotor compared to the nominal power has high production in relation to its nominal power, and a turbine with a small rotor and a large generator will produce much in relation to its swept area. A very high value on one of these key figures can be obtained by choosing a poor relationship between rotor size and generator size. A good turbine should have good values on both key figures.

Capacity factor is a third key figure: it is the mean value of the power of a turbine during a year, compared to its nominal power. The same measure is sometimes expressed as *full load hours*. If the capacity factor is 0.3 this corresponds to 0.3 × 8,760 hours: 2,628 full load hours.

These key figures say more about wind resources than about the efficiency of the turbines.

The most useful key figure is the economic one; total investment [crowns, dollars, euros etc.]/kWh a year. This figure sets the total investment in relation to the production at the site, so it is a useful tool to compare different models, sizes and configurations at a specific site (or some comparable sites).

Box 3.6 Key figures for wind turbines

Power/swept area: $\frac{\text{Annual production}}{\text{Rotor swept area}}$ kWh/m^2

Power/nominal power: $\frac{\text{Annual production}}{\text{Nominal power}}$ kWh/kW

Capacity factor: $\frac{\text{Annual production}}{\text{Nominal power} \times 8{,}760}$ per cent

Full load hours: $\frac{\text{Annual production}}{\text{Nominal power}}$ hours

Cost efficiency: $\frac{\text{Investment cost}}{\text{Production a year}}$ cost/kWh/year

Availability: $\frac{8{,}760 - \text{hours stopped}}{8{,}760}$ per cent

However, since the costs for maintenance are not accounted for, it does not give a final answer to what the most economic option will be.

The conversion efficiency for a wind turbine is not so important. The fuel, the wind, is abundant and free. It is the cost efficiency that matters. It is not reasonable to compare the efficiency of a wind turbine with that of conventional power plants.

Finally there is a key figure for the technical reliability of a wind turbine: *availability*. This figure is given as a percentage. If the wind turbine is out of operation due to faults and ordinary servicing and maintenance for five days in a year, the technical availability is 98.6 per cent. This means that the turbine could produce power in 98.6 per cent of the time, if there is wind enough to make it run.

The technical availability of wind turbines on line is in most cases very high: 95–97 per cent. However, component failures do occur when turbines get older. A large proportion of the stopped time for wind turbines is not caused by faults in the turbine itself, but by unpredicted loss of power in the grid.

The technical lifetime for a turbine is estimated to be 20–25 years. The economic lifetime can be shorter, if the cost of maintenance increases too much as the turbine gets old. A turbine at a good site can ‘pay back’ the energy that has been used to manufacture the turbine in 3–6 months; a very good energy balance compared to other power plants.

However, the technical availability of mass-produced wind turbines is very high, and the technical lifetime is estimated to be 20–25 years. When a turbine has served its time, it can be dismantled and most of the components can be recycled.

References

Claesson, P. (1987) *Vindkraft i Sverige: teknik, tillämpningar, erfarenheter*. Kristinehamn: SERO
EMD (2014) *WindPRO2: Wind Turbine Catalogue*. Ålborg: Energi- og Miljödata.
Gipe, P. (1993) *Windpower for home and business*. Post Mills, VT: Chelsea Green.
Södergård, B. (1990) *Vindkraftboken*. Stockholm: Svensk Byggtjänst.
Stiesdal, H. (2000) *25 års teknologiudvikling for vindkraft – og et forsigtigt bud på fremtiden*. Knebel: Naturlig Energi månedsmagasin.

4 Wind power and society

The preconditions for the development of wind power are set by the national policies of a country and by the laws and regulations that are inaugurated by its parliament. And politicians also decide the economic rules of the game: taxes, charges, subsidies and other means of control often affect energy prices, on top of the actual cost of power production. There are also different ways to promote wind power and other renewable energy sources. In short, politicians set the framework for wind power development and thus influence the pace of development.

Each country has it own laws prescribing the procedures and permissions that are necessary to get permission to install wind turbines, and setting the framework for the economic conditions for the sale and distribution of wind-generated electricity. I am familiar with the procedures and the economic framework applied in Sweden and some other European countries, but it is impossible for me to cover the situation in all countries. Therefore here I try to give a general view, with examples mainly from Scandinavia and other parts of Europe, and hope that what I say applies to most other countries as well.

To install a wind turbine it is in most cases mandatory to obtain building permission from the municipality, also in some cases from the regional authorities – counties and equivalent – or even from the national government. The authorities will assess if the applications are compatible with laws and regulations.

All countries have some kind of energy policy which has been formulated by the government and approved by the parliament.

At the international level there are also agreements, directives and treaties concerning energy. In 2001, for example, the EU adopted a new directive on renewable energy with recommended targets for its member countries. The Kyoto Protocol, which was ratified in 2005, also set obligations for industrialized countries to reduce their emissions of greenhouse gases, and development of wind power is considered to be an effective means to achieve that end. These kinds of international treaties have an impact on the energy policy of specific countries.

Table 4.1 Project size and EIA requirements in Sweden, Denmark and Germany

Country	*Local municipality*	*Regional authority – EIA*
Sweden	1 wind turbine* and 1–6 wind turbines less than 120 m total height	2 wind turbines more than 150 m total height, more than 6 wind turbines
Denmark	1–3 turbines Less than 80 m hub height	4 or more turbines or more than 80 m hub height
Germany	1–2 turbines	3 or more turbines or more than 10 MW**

*Any total height
**Can be required, and is mandatory for more than 20 turbines

Permission inquiry

The *municipality* (or council – terms will of course vary from country to country) takes up a position regarding planned wind power projects by evaluating applications for building permission. The building committee or similar institutions usually have this task, and the members of the board of the committee are local politicians that represent the inhabitants of the municipality. The decisions taken have to conform to relevant laws, in this case usually the building law. The *county administration* in most countries is a state authority at the regional level. For larger projects, this may be the level at which the decisions on permissions to develop wind power projects have to be taken. The county administration undertakes a legal inquiry to see if the project conforms to the relevant laws. In other words, all wind power projects will need permission from the municipality; large projects may need an additional permit from the regional or even the national authorities.

Environment impact assessment

For larger projects it is compulsory to make an environment impact assessment (EIA). The rules which determine the size of project which needs a comprehensive EIA differ from country to country are shown Table 4.1.

The processing of applications to get permission for wind power development differs according the laws and procedures in different countries. In Sweden this process usually takes at least one year, and in many cases several years if there are appeals against the decisions made by the authorities.

Opposing interests

When an application is processed and evaluated, all possible impacts and conflicts with other opposing interests are investigated. The authorities therefore have to refer the application to a number of other authorities for consideration. The most common opposing interests are:

- *Neighbours.* Impacts from noise and rotating shadows from wind turbines can be annoying for neighbours if the turbine(s) are installed too close or in an unsuitable direction in relation to dwellings or holiday cottages.
- *Defence.* Wind turbines can interfere with military establishments for radar surveillance, radio communication etc. In some areas the air force will have objections against high structures.
- *Telecom systems.* There is a risk that wind turbines can interfere with radio, TV and telecom signals and civilian radar.
- *Safety.* To reduce the risk of accidents – if ice is thrown from the rotor blades, a blade or other parts fall off, if the turbine falls over and other incidents – a safety zone may be necessary.
- *Civil aviation.* For civil aviation there are very well defined and strict rules on the minimum distance and maximum height of wind turbines and other structures in the areas surrounding an airport.
- *Protected areas.* In most countries areas that are especially valuable for nature, cultural heritage, recreation or some other common purpose are protected according to the environment laws and similar laws and regulations.

Permission process

The permission processes in different countries are quite similar, at least in Europe. However, in some countries the time from application to permission is quite short, while in other countries it can take years. Another difference is the likelihood for a positive outcome, which also differs very much. Even if the requirements are similar for building permission and the EIA that has to be approved, the efficiency of the process itself differs.

The lead-time for a wind turbine installation is only three months. If all processes move on according to schedule, it should not take more than six months from application until the permissions to build are granted. The time it takes to realize a project is also a political matter. It is a matter of priority and planning.

In Denmark, where development on land has been very fast during the 1990s, the government commissioned municipalities and counties to find suitable areas for wind turbines within their regions. During this planning process much of the assessment in regard to conflicting interests were made. By this *positive* planning, there were only some practical details left for discussion when applications were submitted. The Danes have applied a similar method for their offshore developments, by creating *wind fields* at sea for this purpose.

In many countries such planning has not taken place. It is up to the developer to find a site and then hope that the authorities agree that it is suitable for this purpose.

In the United Kingdom, the British government found an effective solution for offshore developments and created a 'one-stop shop'. An office was created where all aspects and permissions were processed in close dialogue with the developer. The time from application until the wind power plant was on line was radically reduced and the proportion of approved applications increased from 56 per cent in 2000 to 96 per cent in 2003. In Spain a similar method has been used on the regional level. To reduce the time for the application process is a matter of political will and administrative skill.

Wind power politics

In Europe the development of commercial wind power started in the early 1980s. Today countries like Denmark, Germany and Spain have several thousands of MW wind power on-line as well as a new industry with hundreds of thousand of employees. In 2012 there were 249,000 people employed in the wind energy sector in the European Union, according to European Wind Energy Association.

In other countries, development has been considerably slower. It is not very difficult to find explanations for the large differences in the pace of development between different countries. The explanation can be found in the wind power policy that has been conducted by different governments.

Today it is perfectly clear which kind of political measures, rules and regulations promote a fast development of wind power, and what measures can be used to keep development on a slow and low level.

In Denmark, Germany and Spain it has been profitable to invest in, own and operate wind turbines. The laws and regulation in these countries guarantee a fixed and rather high price for the power produced, during the time period that it takes for the owners to earn their investment back. There have been clear political signals that wind power should be developed quickly, and it has not been too difficult or time-consuming to get the permissions necessary to install wind power plants and connect them to the power grid.

The politicians in these countries have had several different motives for their support for wind power. Renewable energy from wind power will reduce the impact from power production on the environment, as well as the emissions of carbon dioxide that most countries have been obliged to reduce in accordance with the Kyoto protocol. The development of a new manufacturing industry creates much new employment and economic growth. Wind power has enabled the politicians to attain these goals.

National energy policies

The energy policy of a country governs the conditions for wind power and its ability to compete in the electric power market. If the energy policy of Denmark, where wind power has been a success story, and Sweden, where development still has not taken off, are compared, the differences are easy to identify. In the paper ‘Possibility of wind power: comparison of Sweden and Denmark' (Miyamoto, 2000) the differences in the energy policy of these two neighbouring Scandinavian countries are described and analysed.

In Denmark, which has a partly decentralized power system, the government has had a strong and explicit ambition to develop wind power. A large share of the electric power in Denmark is produced by coal- and gas-fired power plants. In the 1980s generous investment subsidies were available to farmers and other private investors in wind power, who also could get favourable loans since the financial sector did not see any political risk in such credits. Counties and municipalities were given the task of creating physical space for wind turbines, by making local and regional wind power plans, which had a positive impact on investments.

At the same time the government ordered the utilities to take an active part in the development of wind power and to pay a good price for wind-generated electric power from independent power producers. A very strong domestic market was developed for the wind turbine manufacturers, who also were helped by the government to establish themselves in foreign markets.

Denmark has conducted a long-term wind power policy where economic means of control, planning and other measures have been *coordinated.* By this policy Denmark has managed to adapt its power production to environmental demands, and the obligation to the Kyoto protocol, to reduce emissions of greenhouse gases that were not on the agenda when this development started. At the same time Denmark managed to develop a new industry that soon was dominating the world market (see Figure 4.1).

In Sweden, that has a strongly centralized power system, the government has not had any great ambitions to develop wind power. Since a very large share of the power is produced by nuclear and hydropower plants, there has not been any pressure on the government to reduce emissions of greenhouse gases from power production. The financial sector has not received any signals to invest money in wind power, and has had a negative attitude to this. The rules and regulations for permission to develop wind power are more complicated, which combined with the Swedish traditional protection of nature and lack of coordination between authorities on different levels, has made the permissions process very drawn-out.

The power industry has been passive and the prices for wind-generated power have been very unstable and unpredictable. Even in Sweden private owners have, with the help of independent project developers, installed wind turbines with some subsidies from the state. The development has, however, been too modest to create a strong domestic market and a wind turbine manufacturing industry (see Figure 4.2).

By the wind power law introduced in 2006, the Swedish government for the first time gave a signal that wind power should be developed quickly in Sweden. The reform of the rules for permission applications did not, however, speed up the process, but rather slowed it down. The municipal

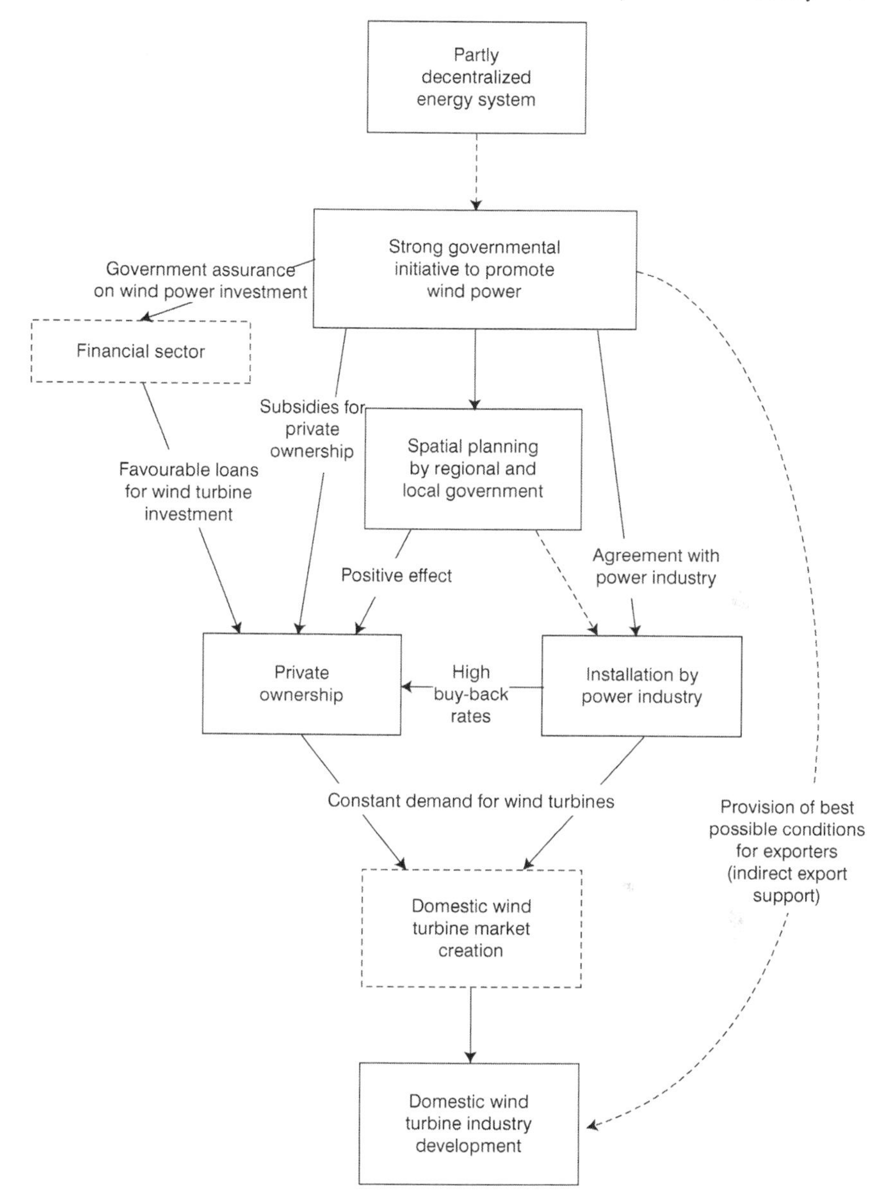

Figure 4.1 Denmark's wind power policy (source: Miyamoto, 2000)

veto (i.e. a rule that the municipality has to give its permission to all wind power projects within its area) has delayed or stopped many projects. After some years, however, the certificate market had some effect, when certificate prices rose to a level high enough to stimulate investments, and the development also took off in Sweden.

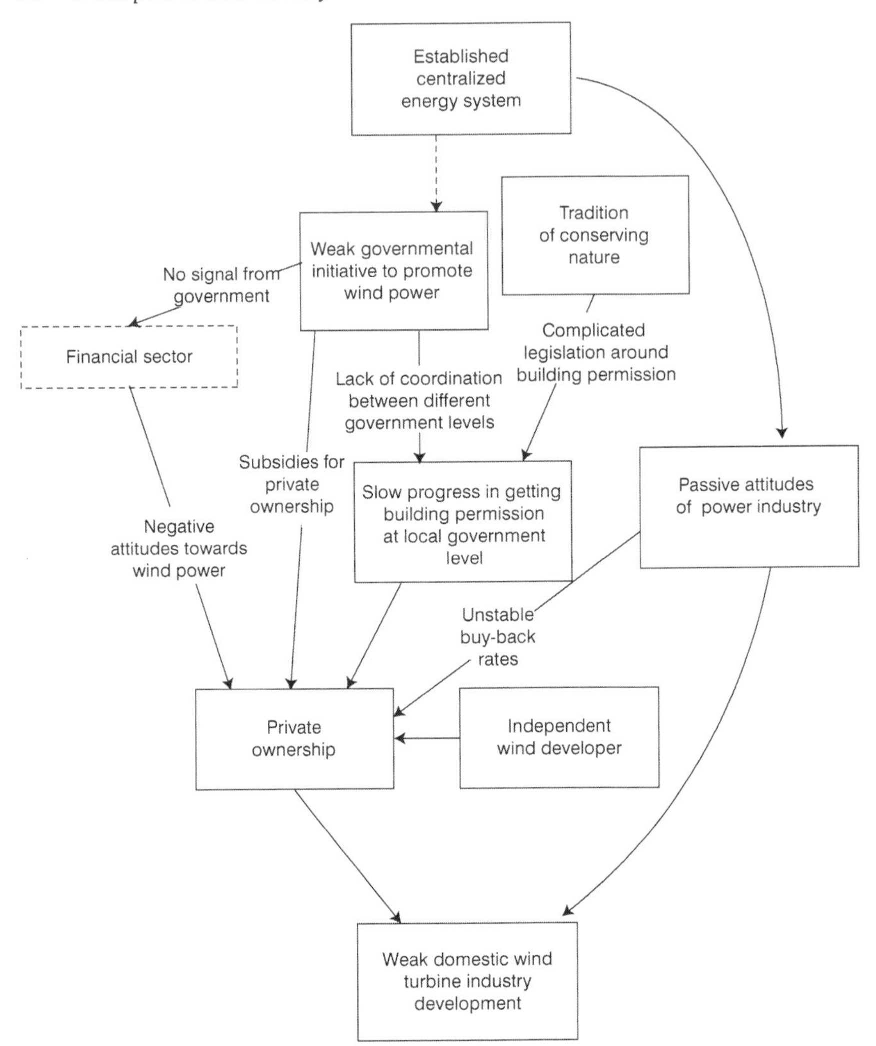

Figure 4.2 Sweden's wind power policy (source: Miyamoto, 2000)

Since the diagram describing Swedish wind power policy was made (see Figure 4.1), the Swedish government has taken some initiatives to speed up the development of wind power in Sweden, but it has not led to the creation of a domestic wind power industry (Swedish industry does, however, manufacture many wind turbine components), the cooperation between different state agencies is still poor and the power industry, especially the state owned power company Vattenfall, is still passive. It has, however, become easier to finance wind power projects during periods with high power and certificate prices.

Support schemes for renewable energy

It is always more expensive to produce electric power by new power plants than with plants that have been on-line for many years. The reason for this is not that new power plants are more expensive, but that older power plants (that often had significant economic subsidies when they were built) already have paid back a large share of their loans and consequently have much lower capital costs. This situation, which makes it necessary to give extra economic support to new power plants, does not apply only to wind power, but to all new power plants.

To make it possible to build new power plants which can replace old ones when they are taken out of operation, some kind of economic support is necessary.

In the RE-XPANSION project, researchers from several European countries have described and evaluated the different support schemes that have been used to promote wind power and other renewable energy sources. The result of their comparative analysis is presented in the report *Support Schemes for Renewable Energy* (EWEA, 2005).

Incentives used to promote wind power can be grouped into three categories:

- *Green marketing.* Voluntary systems where the market determines the price and the quantity of renewable energy.
- *Fixed prices.* Systems where the government dictates the electricity prices paid to the producer and lets the market determine the quantity.
- *Quotas.* Systems where the government dictates the quantity of renewable electricity and leaves it to the market to determine the price.

The first of these, green marketing, has proved inefficient. In surveys quite a large proportion of consumers claim that they would pay a little bit more for power from renewable energy sources like wind power, but when they have this opportunity less than 1 per cent actually choose this option.

Systems with fixed prices or quotas are regulated by law, and are thus compulsory, which makes them more efficient. These can then be divided further and the report defines five different types of support schemes for renewable energy:

- investment subsidies;
- fixed feed-in tariffs;
- fixed premium systems;
- tendering systems;
- tradable green certificate systems.

Investment subsidies

Investment subsidies have been used by many countries in the early stages of wind power development, for example Denmark, Germany and Sweden, and have proved to be quite efficient. The advantages are that it is a simple system and that the subsidies are paid up front. The support cannot be reduced or withdrawn during the lifetime of the project and thus gives security to the investor. The drawbacks are that it does not differentiate good projects from bad. Investment subsidies have now been abandoned in most countries, but were reintroduced for a couple of years in the United States after the financial crisis in 2008/09.

Feed-in tariffs

Fixed feed-in tariffs have so far been the most efficient way to promote wind power. The price paid for wind-generated electric power is fixed, either to a specific value or in relation to the consumer price. The price can be fixed during the lifetime of the turbine, or until a specified target is reached. The system always guarantees that the investors will get their money back. It has been used by the three countries that have been most successful when it comes to wind power development so far: Denmark, Germany and Spain.

Fixed premiums

Fixed premium systems have been used in combination with other promotion strategies. The premiums give a bonus for the avoided costs on health and environment. In practice the value of the premium is set in relation to the power price to make wind power competitive.

Tenders

Tendering systems have been used in the United Kingdom, in the so-called non-fossil fuel obligation (NFFO) system. It didn't work out too well, since many of the projects that won the tenders were never built. It has now been replaced by a green certificate system. Tendering is, however, used for offshore developments in Denmark and the United Kingdom.

Certificates

Tradable green certificate systems for renewable energy have been introduced in Italy (2002), United Kingdom (2002), Belgium (2002) and Sweden (2003). Producers will get certificates in relation to the power produced during a year, and these certificates can then be traded on a certificate market where the price is set by supply and demand.

Evaluation

In the report *Support Schemes for Renewable Energy* these different support schemes have been evaluated through a survey of more than 500 experts from the energy field. According to this survey the most important properties of support systems are *investor confidence* and *effectiveness*. The five support schemes described above have been specified in two versions, one generic and one advanced, where rules that will make the systems more efficient have been added.

In this survey the feed-in tariff system got the top score, with green certificate and tendering at the bottom. Both the generic and the advanced version of the green certificate system got low scores for the most important criteria, investor confidence and effectiveness, compared to other support schemes. The main result of the survey is that the feed-in tariff system is the preferred support scheme by the respondents.

The rules and regulations for these different systems vary, so each country actually has its own unique system. This makes it difficult to compare and evaluate the efficiency of the different systems. From a historical point of view it seems obvious that the fixed price systems are most efficient, since it has made development in Denmark, Germany and Spain so successful. However, there are other countries that have used the same system with less success, such as Greece and France. The support system is but one of several factors that has to be right. According to the authors of the report, there are four main ingredients in a potentially effective overall promotion strategy for renewables:

1. well-designed payment mechanisms
2. grid access and strategic development of grids
3. appropriate administrative procedures and streamlined application processes
4. public acceptance.

The systems with tradable green certificates have now been used for some years. It has proved efficient, as long as certificate prices are high, but also has a high political risk. Politicians, at least in Sweden, have not been prepared to increase quotas to keep certificate prices on a level that ensures early investors make an adequate return. There have been quite strong arguments for a change from the fixed price system to the tradable green certificate system after the deregulation of the market. The quota system is claimed to be more market oriented than the fixed price system. The authors of this report have a different opinion:

> A system where the government fixes quantity and leaves it to the market to determine the price is unlikely to be more 'market oriented' than a system where the government fixes the price and leaves it to the market to determine the quantity.

> The main difference between quota based systems and price based systems is that the former introduces competition between the electricity producers (e.g. wind turbine operators). Competition between manufacturers of plant (e.g. wind turbines), which is crucial in order to bring down production costs, is present if government dictates either prices or quantities.
>
> (EWEA, 2005: 27)

It is possible to achieve a specific target (MW of new power) in a specific time, by giving a guaranteed price for new power produced that will stimulate the necessary investments. This is exactly what Denmark has done with wind power. In Denmark development of wind power actually has reached the goals that have been set several years ahead of schedule. The target set for 2005 was reached already in 1999, and the target to get 50 per cent of the electric power from the wind in 2030 is already within reach. In 2013 wind power penetration in Denmark was 33 per cent.

Independent power producers

In most countries wind power has not been developed by the power companies from the dominant energy system, but by farmers, economic associations and small limited companies formed to install and operate one or small groups of wind turbines, so called independent power producers (IPPs). During the last few years some of the established power companies have entered the arena, mainly to invest in large offshore projects. Even in Denmark and Germany, with a very large share of wind power in the power system, most of the turbines are owned and operated by IPPs. In Denmark there are around 150,000 different owners of wind turbines,;many of them are private households which are members of wind power cooperatives and farmers. In Sweden such independent power producers owned 76 per cent of the installed capacity in 2012 (see Figure 4.3).

In Germany, Denmark and Sweden the dominant power companies and grid operators have been opponents of wind power and still are to some extent. Only during the last few years have these big corporations changed their attitude. Danish Elsam, which now has changed its name to Dong, which developed the large offshore wind farm in Denmark, Horns Rev, changed its mind about wind power by the end of the 1990s. They realized the business opportunities their experience of wind power held and started their own wind power department that now develops large offshore wind power plants in the United Kingdom and other parts of the world. The Swedish state-owned power company Vattenfall bought 60 per cent of Horns Rev in 2005, and thus became a large actor in the wind power business.

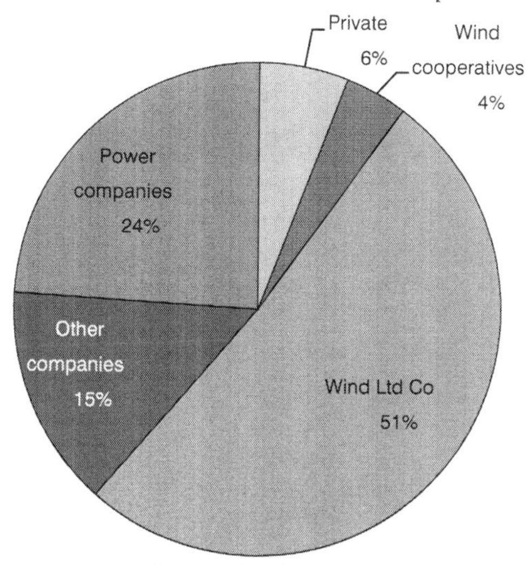

Figure 4.3 Wind power ownership in Sweden 2012. Power companies only owned 24 per cent of the installed wind power capacity in Sweden in 2012. Most of the wind power plants were owned by private individuals, wind power cooperatives, wind limited companies (most of them local) and companies with other business than power production

Competition on equal terms

When the cost of producing wind-generated electric power is compared with the cost of power from other energy sources, the factors that are used for the analysis should be equal. The external cost should also be included, since damage to health and environment by power production generates costs for society.

Power prices do not include the external costs. In 2012 the European Commission updated calculations of the external costs of different energy sources in the report *ExternE: Externalities of Energy* (ExternE, 2012). One of the conclusions is that the costs of producing electric power with coal or oil would double, and increase by 30 per cent for fossil gas, if the external costs for impacts on health and environment were included in the power prices (see Figure 4.4).

If external costs are included in the economic calculation, wind power (with the conditions described above) was already the cheapest means of producing electric power with new power plants in the early 2000s. Since the wind is free, there is no risk that the prices will increase due to changes in the world market prices for fuels, which can have a large impact on the costs for power generated by fossil fuels or nuclear power.

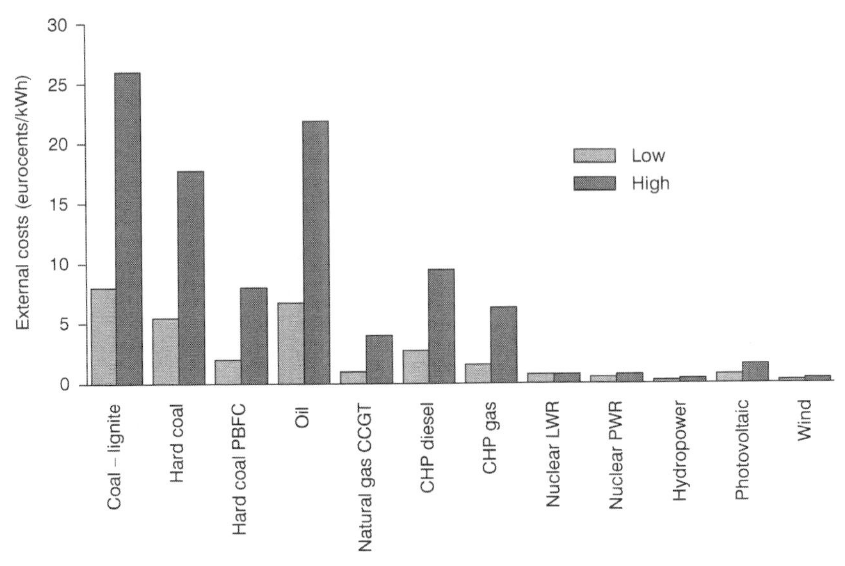

Figure 4.4 External costs. External costs for different energy sources in the EU in 2005, updated 2012 (source: European Environment Agency)

Policy recommendations

Wind power is a relatively new power source that differs from conventional technologies in several respects. Wind power plants are favourable for health and environment, since they don't have any hazardous emissions. Wind turbines are connected to the distribution grid, and can be used for distributed generation. They are modular, that is consist of small units, compared to conventional power plants, and can be added to the power system continuously. Finally, they are often owned and operated by IPPs.

To be able to compete on equal terms on the electricity power markets, that have been created for conventional power plants and are dominated by traditional, often very powerful, power companies, it is necessary, with political support, to eliminate the competitive disadvantage that has been created by decades of financial and political support of conventional technologies. What kind of reforms that are necessary to accomplish this has been analysed and from the results policy recommendations have been formulated (see Box 4.1). Although these recommendations date from 2005, most of the suggested reforms have still not been implemented.

The report *Wind Force 12*, written by Greenpeace for the Global Wind Energy Council (Greenpeace, 2005), calls for *legally binding targets for renewable energy*. Such targets will force governments to develop financial frameworks, grid access regulation, planning and administrative procedures.

Such targets have to be accompanied by policies that eliminate market barriers and attract investment capital. The market has to be clearly defined

Box 4.1 Policy recommendations from RE-XPANSION

The report *Support Schemes for Renewable Energy* (EWEA, 2005) formulates a number of policy recommendations:

- Known external costs should be internalized through appropriate pollution taxes on the polluting energy technologies. The costs of climate change are impossible to quantify, so this should be dealt with in a quantity regime which can assure the adherence to safe maximum concentrations of greenhouse gases in the atmosphere (i.e. by international treaties like the Kyoto protocol, etc.).
- If external costs cannot be internalized by emission taxes, feed-in tariffs or premiums should be used to internalize the differences in external costs between conventional and renewable energy technologies.
- As long as the reduction targets for greenhouse gases are far from securing safe maximum greenhouse gas concentrations, additional measures should be taken to level the competitive playing field for renewables.
- Other reasons for promotion of renewables, security of supply, diversity of supply, local employment etc. should never be overlooked. Such factors need to be taken into account in the making of renewable energy policies in their own right.

in national laws and include stable, long-term fiscal measures, which minimize investor risk and ensure an adequate return on investment.

Further, the report calls for an electricity market reform to remove barriers to renewables and market distortions.

The reforms needed to address market barriers to renewables include:

- streamlined and uniform planning procedures and permitting systems, and integrated least cost network planning;
- access to the grid at fair, transparent prices and removal of discriminatory access and transmission tariffs;
- fair and transparent pricing for power throughout a network, with recognition and remuneration for the benefits of embedded generation;
- unbundling of utilities into separate generation and distribution companies;
- the costs of grid infrastructure development and reinforcement must be carried by the grid management authority rather than individual renewable energy projects;
- disclosure of fuel mix and environmental impact to end-users to enable consumers to make an informed choice of power source.

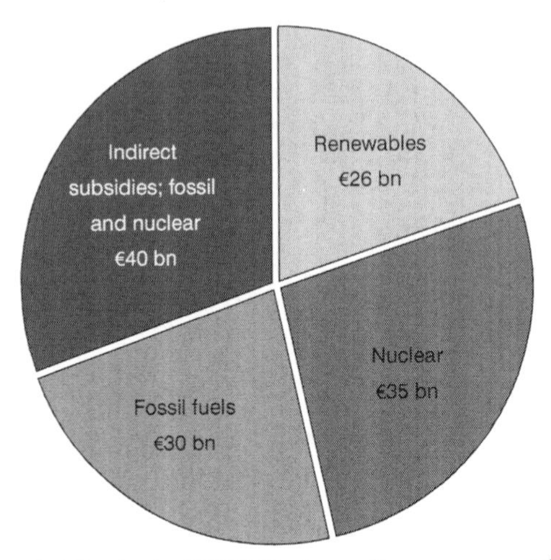

Figure 4.5 Energy subsidies in the EU 2011. According to a report from the European Commission (2013), nuclear power was subsidized by €35 billion and fossil fuels by €30 billion in 2011. The indirect subsidies were €40 billion. These indirect subsidies were, according to the EC, costs for impacts on health from coal-fired power plants, etc. Subsidies to renewables were €26 billion euros (source: based on data from *Süddeutsche Zeitung*, www.suddeutsche.de)

An important point is that fossil fuel and nuclear power sources still receive large subsidies, which distort the markets and increase the need to support renewables (see Figure 4.5). Wind power would not need special provisions if markets were not distorted by the fact that it is still virtually free for electricity producers to pollute, according to the Greenpeace report.

Wind power and environment

Wind turbines use the renewable power from the wind; they don't have any emissions and don't require any fuel transport which can harm the environment. A wind turbine pays back the energy that has been used to manufacture it in three to nine months; the time depending on the wind resources at the site, the size of turbine and the method of calculation. The turbine can be dismantled without leaving any lasting traces behind, and most of the material can be recycled. All other ways of producing new electric power have greater impacts on the environment.

From an environmental point of view, wind power is the best option, and has a positive impact on the global and regional environment. The risks for damages from climate change, acidification, eutrophication and the impacts from this on agriculture, forests, lakes, landscape and human health decreases (see Table 4.2).

The concept of environmental impact encompasses a lot of different kinds of impacts. Wind turbines can cause impacts on the environment by

Table 4.2 Environmental impact from different energy sources

Energy source	*Raw product*	*Emission*	*Other impacts*
Combustion	Coal, oil, gas	CO_2, NO_x, SO_x, VOC, ash	Oil exploitation, mines, transport
Combustion	Biomass	NO_x, SO_x, VOC, ash	Forestry, transport
Hydro power	Running water	None	Exploitation of land and watersheds
Wind power	Wind	None	Land use, noise
Solar heating, PV-cells	Solar radiation	None	Land use

CO_2 = carbon dioxide, NO_x = nitrogen oxides, SO_x = sulphur oxides, VOC = volatile organic compounds

noise, shadow flicker, visual impact on the landscape, flora and fauna, and nature and cultural heritage. Another impact is that the emissions from the power system are reduced. To be able to do reasonable assessments of these impacts, this concept has to be specified (see Box 4.2).

Impacts on the environment can be global, regional or local. Burning of fossil fuels (coal, oil, fossil gas) gives emissions of the greenhouse gases carbon dioxide, sulphur oxides and nitrogen oxides, volatile organic compounds (hydrocarbonates, etc.), heavy metals (lead, cadmium, mercury) as well as soot and particles. The exploitation of fuel from mines, oil and gas wells, causes serious local impacts on the environment, and in that process there are emissions as well. The transport of the fuel from the source to the power plants requires energy and creates its own emissions.

Box 4.2 Environmental impact from wind turbines

The impact on the environment can be divided into the following categories:

Ecosystem:

- chemical/physical impact – acidification, eutrophication, climate change, pollutants, etc.
- impacts on flora and fauna (nature)

Health and well-being:

- impacts on neighbours that can cause nuisance – noise, shadow flicker, safety

Culture:

- visual impact on landscape
- cultural heritage.

Table 4.3 The contribution of wind power to reduction of environmental impact in the Nordic power system

Wind power production	*1 kWh*	*1 GWh*
Annual reduction of emissions		
Sulphur dioxide	0.11 g	110 kg
Carbon dioxide	784 g	784,000 kg
Nitric oxides	0.23 g	230 kg
Particulates	0.02 g	20 kg

The environmental gain that development of wind power creates depends on the power system where the turbines are installed; how the power would be produced without wind power and how that would affect the environment.

Sweden for example belongs to a large north European power system. The power systems in the Nordic countries Sweden, Norway, Denmark and Finland are interconnected and there are also cables for power exchange with Germany, Poland and the Netherlands. When a new wind turbine starts to produce power in Sweden, it replaces the same amount of power from a coal-fired plant in the north European power system. New wind turbines in Sweden do not replace power from hydropower stations and nuclear reactors, but imported power from a coal-fired plant (Holttinen, 2004). By comparing with the emissions from coal-fired power plants, the environmental gains can be quantified (see Table 4.3).

The annual gains for the environment from a wind power project are calculated by multiplying the estimated annual power production with the figures in the Table 4.3. To this, solid refuse in the form of waste (clinker) with approximately 53 g/kWh (53 ton/GWh) is added. (Danish Wind Turbine Owners Association, 2011).

The amount of reduction depends on what power plant is used as comparison in the calculation. CO_2 emissions depend on the efficiency of coal-fired power plants, and emissions of SO_x and NO_x however can vary. Table 4.3 refers to a coal-fired power plant with good equipment for emissions reduction.

The amount of reduction also depends on what type of power plants are installed in the power system where wind power is connected, and the actual share of wind power of the power production – the wind power *penetration*. When the penetration increases, wind power will replace not only coal-fired power plants, but also some hydropower and nuclear power plants in the Nordic power system. With a 4.3 per cent penetration (16 TWh/a) CO_2 reductions are reduced to 700 g/kWh (produced by wind power); with 12.2 per cent (46 TWh), reductions are 650 g/kWh (Holttinen, 2004).

In the Baltic power system, where a large proportion of the power comes from oil-shale-fired power plants, the reductions are larger, 1.05 kg/kWh.

However, to get an actual reduction of the CO_2 emissions, the penetration has be quite large, since the fossil power plants only can be regulated in steps of 10 MW.

The amount of reductions of CO_2 and other emissions depends on the design and characteristics of the power system where the wind turbines are installed, and these figures have to be calculated separately for each power system. In most power systems wind power will give a significant reduction of CO_2 emissions.

Development of wind power contributes to reduce the negative impact from power production on global environment, since carbon dioxide emissions that can change the global climate are reduced. It contributes to reduce emissions of cross-boundary air pollution from sulphur- and nitrogen oxides that cause acidification, eutrophication and other environment impact on the regional level. Wind power has a *positive* impact on the global and regional environment.

Land demand

It seems obvious that wind power needs more land than other energy sources. If you analyse this question and compare other kinds of power plants, it turns out to be less self-evident. Power plants which use fossil fuels or uranium use land areas in the whole chain of production, from the exploitation of the raw material to waste dumps, mines, oil wells, refineries, ports and storage facilities. The demand for land for wind power varies from 0.018 to 0.49 ha/MW including foundations, access roads, transformers and other equipment, according to an empirical study. A British nuclear plant demands 16 ha/MW, according to the same study.

These comparisons of land demand from different power sources were made in the late 1990s. Since then the size of wind turbines has increased very much, and this reduces the land demand of wind power.

Land demand defined as the area limited by a line surrounding the outer towers in a group of turbines depends on the configuration. If the turbines are sited along a line, the land required is very small; if they are arranged in three rows with four turbines in each row, the land required increases. A group of twelve 1.5 MW turbines in a 3 × 4 group need an area of 81 ha, if they are sited in a 2 × 6 group they only need 47 ha. The total nominal power of such a group is 18 MW, which gives a land demand of 4.5 and 2.6 ha/MW respectively. However, at least 95 per cent of this area can still be used as before. The actual land need is then only 0.225 and 0.13 ha/MW.

If the same nominal power is installed using 3 MW wind turbines, only six wind turbines are necessary, but the distance between the turbines has to be greater. Six wind turbines (with 90 metre rotor diameter) in two rows with three turbines in each row require an area of 56 ha. Four wind turbines with a nominal power of 5 MW (with 128 metre rotor diameter), can be sited in a square, and require an area of 57 ha. These larger turbines need about the same land area per MW, but they have higher towers and produce more, so power production per land area will increase with the size of the wind turbines. The better wind resources that are used, the fewer turbines have to be installed.

Local impacts

Access roads have to be built to make it possible to transport the turbines to the site, and a work area for the cranes has to be cleared. Power cables have to be installed to connect the turbines to each other and to the grid.

Wind turbines are usually mounted on a concrete foundation, but can also be bolted into the ground if it consists of solid rock. A concrete foundation consists of a large, square or circular foundation cast in a mould a few metres below ground level and a concrete pedestal with a steel flange where the tower is bolted. The foundation is covered with earth and the foundation does not cover more area than the tower base – a few square metres at ground level.

An access road has to be built from the closest road so that an excavator, mobile crane and other heavy vehicles can get access to the site. These accesss roads can be temporary, and may simply require strengthening existing roads or laying steel plates on the ground. The requirements of the access road depends on the terrain conditions and the size of the turbines. When the turbines are delivered, the access road has to be able to bear heavy lorries with trailers and a heavy mobile crane.

Single wind turbines are directly connected to the grid by a cable buried in the ground. Groups of wind turbines are connected by an internal grid, and this is then connected to the power grid. Large wind power plants also need a substation.

The *direct physical impact* on the environment consists of foundation, access road and cables to the grid, and the turbine itself that will demand some of the air space at the site. When the turbines have been installed, the land can be used as before, as arable land or pasture.

The wind turbines do not cause any emissions that have impacts on the environment. The rotor creates some noise, the turbines are visible during daytime and when the sun shines the rotor throws a rotating shadow that moves from west to east from sunrise to sunset.

Wind turbines have a technical and economic lifetime of at least 20 years. It can be prolonged by replacement of vital components, such as gearboxes, generators and blades. The foundations have a much longer lifetime, and could eventually be reused for a new installation at the same site. The rapid technical development of turbines has made this less likely, but in the future this may be an option. Wind turbines can be dismantled in a day and the site can be restored to its original state. Most of the components can be recycled. Wind turbines do not make a lasting impact on the environment.

Nature

The impact on flora and fauna depends on the type of vegetation and animal life that occurs in the area.

Flora can be affected during the building phase or by changes in the hydrological conditions due to the foundation, cable ditches etc., but this is seldom a problem. Concerning wildlife, the risk for impacts on birds and bats has been debated, and a lot of research has been done to clarify this issue.

In the 1980s there were problems in Denmark with small birds building nests in the nacelles, so the manufacturers had to cover all openings with lattice. In the United States many falcons collided with the rotor blades. The reason for this turned out to be that the smaller birds that were the falcons' prey built their nests in the towers. These turbines had lattice towers that had many forks in their structure perfect for bird nests. Covering the lattice towers with sheet metal solved this problem. In southern Spain, several vultures were killed in a wind power plant in Tarifa. It turned out that the vultures gathered at a large waste dump in the middle of the wind farm and moving the dump to a better place solved the problem. The risk of birds colliding with wind turbines has by experience been proven to be rather small.

In the Altamont Pass in California, there are still problems with raptors colliding with turbines. Since the large 150 MW wind farm on the island Smöla in Norway with 68 wind turbines in operation was built in 2002, around 40 sea eagles have been killed by collisions with wind turbines. Around 150 sea eagles have their territories on this island and since there are no trees they build their nest on the ground. This has, however, not had any impact on the sea eagle *population* on Smöla, which has increased. Today there are technical systems, like dtBird, which warn and deter eagles and other birds when they fly too close to wind turbines. Such technical systems are now used in wind power plants in Spain, France, Greece, Italy and on Smöla in Norway to prevent bird collisions.

However, birds live a precarious life, and about 30 per cent of the birds die during their first year due to collisions with natural or man-made objects (windows in houses, high structures, power lines etc.). In the United States scientific studies have been made to estimate the number of bird deaths a year from different causes (see Table 4.4).

Several studies have been conducted on bird mortality caused by wind turbines. The National Wind Coordinating Committee in the United States analysed all these studies made up to the end of 2001, and estimated the total annual mortality from 3,500 operating turbines in the US to 6,400 bird fatalities per year for all species combined. This would correspond to 0.01–0.02 per cent of the annual avian collision fatalities caused by manmade structures and activities in the United States (NWCC, 2001). Since then much more wind power has been installed in the United States, but still the risk for birds colliding with wind turbines is rather small.

Table 4.4 Bird mortality from collisions, United States

Object	*Mortality, million birds a year*
Power grid	130–174
Cars and trucks	60–80
Buildings	100–1,000
Telecom towers	40–50
Pesticides	67
Cats, domestic and feral*	39
Wind turbines	0.0064

*Wisconsin only (Sagrillo, 2003)

Another impact could be that birds are scared away from areas where wind turbines are erected. This scaring effect varies for different species, most birds don't seem to be frightened by turbines and get accustomed to them quite soon.

The impact on seafowl from offshore wind power plants has also been investigated thoroughly. Comprehensive studies have been made at Tunö Knob and Nysted in Denmark as well as in the Kalmar Strait in Sweden, where a wind farm has been installed in the middle of the migration route, where 1.5 million birds will pass the 12 wind turbines each year. The routes of the migrating birds have been observed visually in daytime and by radar at night and during mist, from 1999 (before the turbines were installed) to 2003. During this period one collision was observed. The worst case for birds killed by colliding with the wind turbines was estimated to 14 birds a year, which is considered negligible. Hunters get permission from the county administration to kill 3,000 seabirds each year in the area. The study shows that birds will spot and avoid the turbines in all kinds of weather.

Important resting places for migratory birds and nesting areas are usually designated as bird protection areas. However, it is always important to investigate the bird situation in coastal areas and offshore and to adapt wind power plants to these conditions to minimize any impact.

Bats can get killed when hunting insects which gather around wind turbines. Almost all of these accidents (90 per cent) happen during warm nights with slow breezes in late summer and autumn (end of July to September) or (10 per cent) in spring (May to early June).

Some insect species migrate, just like some bird and bat species do. It is during migration that these insects gather around wind turbines, and it is during these migrations that the risk for bat collisions occurs. The risk varies for different bat species. Those who hunt insects on high altitudes in free air are most vulnerable; in Europe 98 per cent of the bats killed belong to such species. Since it is known when and during which weather conditions such accidents can happen, it is also possible to avoid them by turning turbines off during these risk periods. Technical systems for this have also been developed.

Global climate change has severe impacts also on birds. According to the report *Bird Species and Climate Change* from WWF (WWF, 2006), where more than 200 scientific papers have been analysed, global warming will have impacts on birds all over the world. Some species stop migrating, others starve when they change migration routes. Migrating birds and seafowl are worst off, and species that live in mountains, on islands, wetlands and in the Arctic or Antarctic.

A transition of the electric power system to renewable energy will also benefit birds and bats.

Sound propagation

Wind turbines can cause two different types of noise: mechanical noise from the nacelle (gear box, generator and other moving mechanical components) and aerodynamic noise from the rotor blades. There are carefully specified rules and methods for how noise from turbines shall be measured, how manufacturers should specify it, how the noise immission should be calculated at different distances from a turbine and which sound levels should be permitted at different types of buildings.

The main distinction concerning measurements of noise is made between sound *emission* and sound *immission*. The sound emission is the sound that the turbine sends out – emits. The value for sound emission that the manufacturers declare and which are used to calculate sound levels at different distances from the turbine is the sound emission from the centre of the rotor when the wind speed is 8 m/s at 10 m height above ground, for a turbine sited in an open landscape with roughness class 1.5 (roughness length 0.5).

The sound immission is the value that is measured (or calculated) at a specific distance from the turbines. If the sound emission and the hub height are known, the sound immission at different distances from the turbine can be calculated.

Sound is measured in dBA (decibel A), that is an A-weighted sum of the sound with different frequencies, which is adapted to the human sensitivity to sound at different frequencies. The unit dBA gives a measure of the sound that the ear registers. Normal speech has a sound level of 65 dBA, a modern refrigerator 35–40 dBA, a city street about 75 dBA, a discotheque around 100 dBA and a quiet bedroom 30 dBA.

The sound emission from a wind turbine can vary from 95 to 110 dBA. The sound from modern wind turbines comes from the rotor blades; the mechanical noise has been eliminated by sound-absorbing materials in the nacelle, better precision in the manufactured components and through damping. Nowadays, mechanical sound can be heard only when a component begins to fail. The sound from the rotor is an aerodynamic

Table 4.5 Sound level from wind turbines/distances (m)

	Immission		
Emission	*45 dBA*	*40 dBA*	*35 dBA*
105 dBA	350 m	575 m	775 m
100 dBA	200 m	350 m	575 m
95 dBA	120 m	200 m	350 m

swishing sound. Infrasounds from wind turbines are far below the levels people are accustomed to from other sound emissions in their daily life.

Manufacturers have managed to reduce this swishing sound during the years of continuous development, by changing the form of the blades. The level of sound emission from a turbine is decisive for the distance to neighbouring houses (see Table 4.5).

The sound emission from wind turbines (that are found in the technical specifications for the turbines) is usually in the range 95 to 105 dBA. The table shows rounded values to give an idea of the distances and how they differ for various emission values. Decibel uses a logarithmic scale; an increase by 3 dBA corresponds to a doubling of the sound level (power).

The sound from wind turbines differs from other kinds of industrial noise. The aerodynamic swishing has the same character as the rustling of leaves or other sounds induced by the wind. Turbines with variable speed rotate slower in low winds and the noise level is lower than the background sound level at almost all wind speeds.

Wind turbines can only be heard under certain preconditions. When the wind wanes the turbines stop and can't be heard at all. When the wind speed is higher than 8 m/s (at 10 m agl) the sound from the turbine is masked by background sounds from rustling leaves and other wind-induced sound. Wind turbines will be heard only when the wind speed is between start wind speed 3–4 m/s and up to 8 m/s and the recommended values for maximum sound level are reached only at 8 m/s at 10 m height. The sound will spread more on the lee side of the turbine. In other directions the sound level will be lower.

Calculation methods

The sound immission at various distances from a wind turbine can be calculated with models for sound propagation. There is an international standard method for this, ISO9613-2. Unfortunately the authorities in some countries have decided to use their own models instead. Their acoustic experts claim that their models are more accurate for wind turbines than the international standard model. The results from these different models however do not differ very much (see Table 4.6).

The results 500 metres from a wind turbine differ by less than 2 dBA. Considering that the smallest difference in sound level that the human ear

Table 4.6 Sound emission according to different calculation models*

ISO9613-2	*Denmark*	*Holland*	*Sweden*
34.1	35.5	36.0	35.5

*For a wind turbine with 50 m hub height and a sound emission of 100 dBA; sound immission 500 m from the wind turbine (dBA)

Table 4.7 Distance to 35, 40 and 45 dBA*

dBA	*ISO9613-2*	*Denmark*	*Holland*	*Sweden*
35	465	525	555	525
40	305	325	325	325
45	205	195	185	195

*For a wind turbine with 50 m hub height and a sound emission of 100 dBA

can perceive is 3 dBA, the results are in practice the same. For the minimum distance to a dwelling, where the maximum sound immission is 35 dBA, the distance will be a little more than 500 metres for Holland, Denmark and Sweden and slightly less than 500 metres for the international standard, that is used in Germany and other countries. Close to the turbine the international standard is 'stricter' while the difference at 40 dBA is only 20 metres (Wizelius et al., 2005) (see Table 4.7).

Rules and regulations

Rules for recommended sound immission limits at neighbouring houses differ from country to country. In Denmark the limit for dwellings is 45 dBA, in Sweden the limit is 40 dBA. In Britain the sound immission may not increase by more than 5 dBA above the background noise level (see Table 4.8).

Calculations on sound emission from wind turbines are based on the assumption that the ground is hard and flat so that all sound waves are reflected. Vegetation, buildings and other structures that can absorb or deflect sound are not taken into account. It is this worst case that sets the limit for the sound propagation zone when these rather simple linear calculation models are used. For wind power plants with many wind turbines in a large area the calculation results are always too high, since sound emission also depends on the wind direction, and is higher on the leeside of wind turbines, and a building can be leeside of only one turbine at any given time. More advanced calculation models, like Nord2000, consider the difference in sound emission in different directions, the character of the ground and other parameters, and give a more accurate calculation of the sound immissions that actually occur at dwellings in the vicinity. The use of Nord2000 is now also accepted in Sweden and some other countries.

Table 4.8 Recommended limits for sound immission in different countries, dBA

Country	*Work areas – offices, industry*	*Dwellings*	*Villages, farms*	*Recreation areas*
Denmark	–	39, 37[a]	44, 42[a]	40
Germany	50–70	40	45	35
Holland	40	35	30	–
United Kingdom[b]	+ 5	+ 5	+ 5	+ 5
France[b]	+ 3	+ 3	+ 3	+ 3
Norway	40	40	40	40
Sweden	50	40	40	35

a At 8 m/s and 6 m/s respectively. Max indoor low frequency noise max 20 dBA at the same wind speeds
b Maximum increase over background noise during the evening and night

Shadows and reflections

During some periods of the day wind turbines can create shadows that can be disturbing if the turbines are unsuitably sited in relation to neighbouring buildings. Reflections from rotor blades is no problem, since the blades on modern turbines have an anti-reflection coating. The rotating shadows from rotors can create a stroboscope effect when they pass a window, which can become an unpleasant surprise if this risk hasn't been considered before the turbines were installed.

The risks for disturbing shadows are greater the closer a house is to a wind turbine. However, due to rules of maximum sound immission, the minimum distance to the closest neighbouring dwelling usually is 6–10 rotor diameters and then shadows will occur only during some short periods each day in limited periods of the year. Increased hub heights and rotor diameters have, however, increased the size of shadows as well. A shadow will be 'diluted' with distance, so that its sharpness decreases and the shadow finally disappears due to optical phenomena in the atmosphere.

Theoretically a shadow from a turbine can reach 4.8 km (a turbine with a 45 m rotor diameter), which would occur just after sunrise and just before sunset. In reality a shadow will have a maximum reach of 1.4 km (2 MW turbines with 2 m blade width) although shadow effects often are calculated for distances up to 2 km.

The shadow from a wind turbine moves in the same manner as the shadow of a sundial, from west through north and to the east from sunrise to sunset. Since the sun will rise later in the winter, when the sun's altitude also will be lower, the shadow will move along different paths in different seasons. Since the altitude of the sun can be calculated exactly for each time of the day and for different latitudes, the path of the shadow at each place can be exactly calculated (see Figure 4.6).

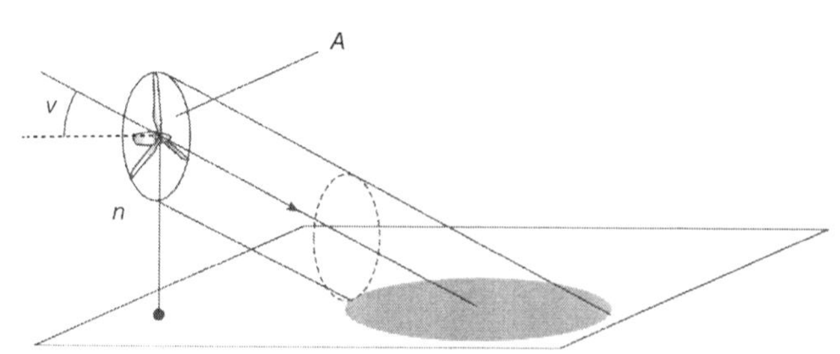

Figure 4.6 Model for shadow calculation. If the hub height (*n*), the rotor area (*A*) and the inclination of the sun in relation to the horizontal plane (*v*), the location of the shadow can be calculated (illustration: EMD/Typoform)

A house situated due west of a wind turbine can get shadow flicker at 6 o'clock in the morning, a house north of a turbine at 12 o'clock, and a house east of a turbine at 6 o'clock in the evening. In daytime the shadow is much shorter in the summer than in the winter; if the houses are 500 metres away from the turbine none of them will get shadow flicker more than during two short periods and for a maximum 20 minutes per day. During which time in the day a shadow from a wind turbine will fall in a specific place and for how many hours a year this may occur can be calculated using a web-based calculator that is available at the Danish wind power industry association guided tour: www.wind powerwiki.dk.

The results of these estimates or calculations show the theoretical maximum times a house can get shadow flicker from the turbine. It shows the 'worst' case, which is if the sun always shines and the wind always blows from a direction that gives maximum shadow impact (rotor perpendicular to the window). Since the sky sometimes is overcast and the wind speed and direction varies, the real time for shadow flicker is much lower, in most cases less than a third of the worst case. If information about percentage of time with sunshine/month and on the distribution of wind speed and wind directions is available, the time for actual shadow impact can be calculated.

The calculation model in Figure 4.6 is however based on some simplifications, which give an overestimation of the shadow impact. In this so-called geometrical model the sun is reduced to a dot, and the light/shadow spreads in vacuum. However, the sun covers 0.5 degrees of the sky, so that sunbeams actually meet behind the rotor blades and wash out the shadows, at a certain distance. The air in the atmosphere also diffuses the light. These two phenomena have been studied by the German scientist Hans-Dieter Freund, who has introduced a physical correction factor to get results that better mirror reality.

The maximum distance for a shadow to be visible depends on the hub height and rotor diameter of the wind turbine. The length of the shadow

Table 4.9 Maximum length of shadows from wind turbines

		Summer		*Winter*	
Hub height	*Rotor diameter*	*Horizontal*	*Vertical*	*Horizontal*	*Vertical*
25	25	200 m	350 m	300 m	700 m
50	50	300 m	700 m	600 m	1,250 m
75	75	500 m	1,100 m	850 m	1,800 m
100	100	600 m	1,375 m	1,100 m	2,300 m
125	120	700 m	1,650 m	1,300 m	2,700 m

also varies with the opacity of the air, which depends on the humidity and temperature. On a clear day in winter the shadow can be longer than on a clear summer's day. The shadow will be visible at longer distances on a vertical than on a horizontal surface. According to Freund the maximum length of shadows can be calculated (see Table 4.9).

The maximum distance for a shadow on a vertical surface, a window or façade, are for wind turbines with 75 metre hub height and the same rotor diameter, 1,100 metres in the summer and some hundred metres longer during the rare winter days when the air is crystal clear. The size of the shadow, the area it sweeps, will also decrease with distance (Freund, 2002).

Visual impact on the landscape

Wind turbines are visible objects and consequently have a visual impact on the landscape, like most other structures: factories, power lines, highways etc. Since wind turbines are tall and have a rotating rotor, they attract the attention of passers-by. Therefore wind turbines are considered to have a comparatively large impact on the impression a landscape gives. Whether this impact should be considered positive or negative is a matter of individual opinion, and also varies in different types of surroundings. Some consider wind turbines to be ugly machines that turn the environment into an industrial area, others view wind turbines as slender sculptures that visualize the power of the wind, or as a clever, and therefore acceptable, way to use the power that nature offers for free.

Farmers usually want to use the available natural resources in an efficient way, and consider the surroundings as a production landscape. They also have to make their living out of it. Tourists and holiday cottage owners often view the landscape as a 'post card' that is there to give aesthetic and similar experiences. Furthermore, opinions and experiences tend to change over time, so that wind turbines after some time are considered as natural and valuable components in the landscape.

It is difficult to determine how wind turbines affect the landscape, since the experience of the values of a landscape is subjective and different people can have quite different opinions about it. There certainly are some turbines that are sited in bad places, as well as turbines that suit the landscape well.

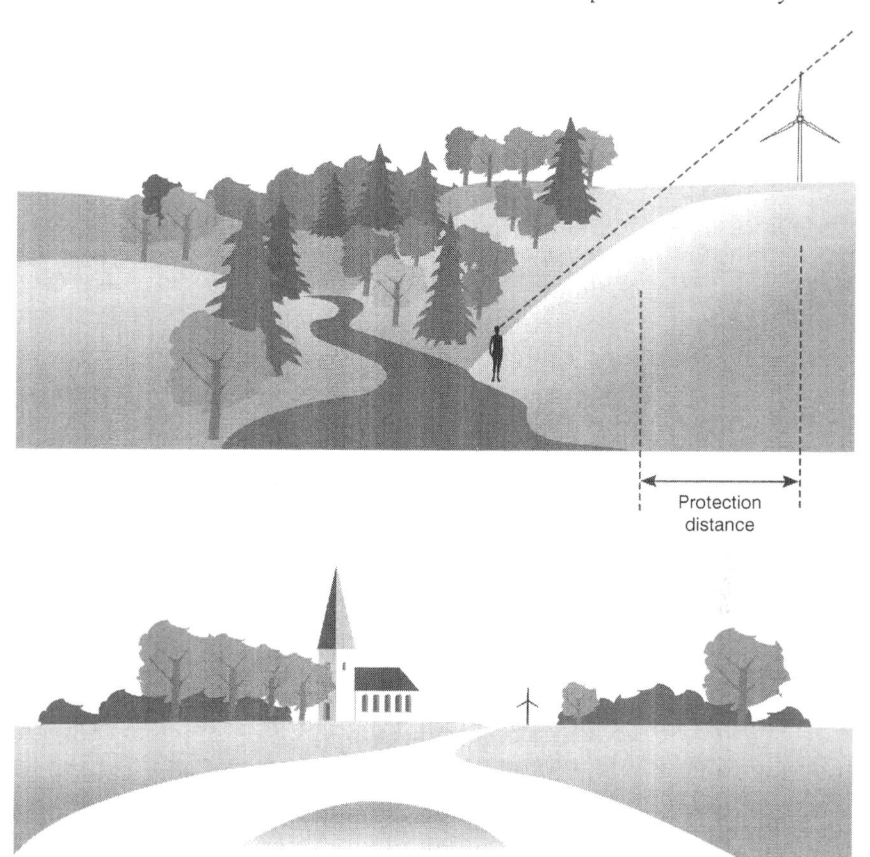

Figure 4.7 Siting of wind turbines in the landscape. It is important to consider from what places a planned wind turbine installation will be visible and what impact the turbines will have on the experience of existing buildings and nature. This can be done by checking the views from different viewpoints. In a valley the brows of hills should be kept free, and distances from churches and similar buildings should be great enough so that the turbine(s) don't destroy the impression of them (illustration: Typoform/Länsstyrelserna i Skåne, 1996)

Obviously there are some factors that make a difference. However, it is very hard to formulate general rules for how turbines should be sited in the best way in the landscape (see Figure 4.7).

The developer who plans to install wind turbines chooses the site very carefully, to make the turbines produce as much power as possible. The area that surrounds the turbine(s) should be as open as possible, with a proper distance to buildings, woods and other obstacles that can disturb the wind. This usually is a good site also from an aesthetic point of view.

The visual impact of wind turbines decreases quickly with distance. A limit for when the visual impact can be considered to be negligible has to be set. A common rule of thumb is that wind turbines dominate the landscape within a

Table 4.10 Visual zones for wind turbines

Total height	*Close zone*	*Middle zone*	*Distant zone*	*Outer zone*
	0–4.5 km radius	*4.5–10 km radius*	*10–16 km radius*	*>14–16 km radius*
150 metre	Dominating	Visibility varies depending on character of the landscape	Clearly visible but not dominating	Small structures on the horizon, hard to detect in some weather conditions

distance of ten times the turbine's hub height, i.e. within a circle with 900 metres radius for a turbine with a 90 metre tower. This will be outside the 40 dBA-limit for sound immission. Further off, the turbine is clearly visible but does not dominate the landscape and after 10–16 km it melts into the landscape.

A wind turbine will be visible up to a distance of 400 times the hub height, i.e. up to 36 km for a turbine with 90 m hub height. To be visible there must be a free view and clear weather. At a distance of 20 km most people will need binoculars to catch sight of a wind turbine (see Table 4.10).

The impact on the landscape of a planned wind power project can be illustrated with photomontages. Most wind power software programs offer this option, but it can also be done with Photoshop or similar software.

The number and size of the turbines and their configuration also influence the visual impact. Experience has shown that it is very hard for the human eye to distinguish between small and large turbines. The eye will interpret a difference in height to a difference in distance. It is also hard to note different patterns in a wind power plant.

It is not possible to assess how the values of a landscape will be affected by new structures or other changes just by reading maps or by interpretation of laws and regulations. These values are defined by the evaluations made by the people who work and live in the landscape. The value of different sites and views depend on the traditions, memories and feeling of the local community. These aspects are usually clarified during the public consultations that are a part of the permissions process.

Finally it is important to realize that the landscapes we live in are created by man. There is no natural landscape. This is the view of modern geographers:

> Landscape is not scenery ... it is really no more than a collection, a system of man-made spaces on the surface of the earth. The natural environment is always artificial.
>
> (J.B. Jackson, quoted in Pasqualetti et al., 2002)

A landscape in a sustainable society will look different from the landscape of today.

Wind power planning

Wind power planning can mean several things and can be undertaken on different levels by various actors. To start at the top level we have international treaties – that are legally binding – or recommendations, like the Kyoto protocol. To achieve the targets for reductions of CO_2 emissions, to address the threat of global climate change, an important step is to replace fossil fuels with renewable energy sources, one of which is wind power. These kinds of international treaties govern overall ambitions for development of wind power and other renewable energy sources on the national and international level.

Most countries have some kind of national energy policy, of which development of wind power can be a part. They could set a target for how much and how fast wind power should be developed on a national and sometimes also regional level. An important part of this policy is the set of laws, rules and regulations that are applied to wind power development. These aspects have been discussed in the preceding section.

Spatial planning at the local and regional level is governed by local and regional authorities, which handle building permissions, environmental permits, etc. In some regions and/or municipalities comprehensive plans have been worked out, where suitable areas for wind power development are designated.

On the other side of the coin there are developers, who make their own plans for wind power development. Companies have some overall business ideas, business plans, strategies and ambitions, which result in surveys for suitable sites for wind power, feasibility studies and project development. This kind of planning is described in Chapter 6 'Project development'.

Here the focus is on the spatial planning. A policy framework – rules and regulations that stimulate or act as barriers to wind power development – regulates this planning. The legislation for planning varies in different countries. A comparison will give some ideas on how to implement an efficient planning strategy and to avoid planning methods that will create barriers.

Targets for wind power development

In Denmark the government has set targets for wind power development, the first in the government energy strategy *Energy 81*, which set the target to 1,000 MW wind energy by the year 2000. This was followed by the action plan *Energy 2000* in 1990 and *Energy 21* in 1996, that set the target to 1,500 MW by 2005, a target that was reached already in 1999. A long-term target for 2030 was also set at 5,500 MW, out of which 4,000 MW were expected to be installed offshore. With this target 50 per cent of the electric power in Denmark will be supplied by wind power.

In 1992 the Wind Turbines Law was enacted. The government required municipal and regional authorities to make plans for wind turbine siting through a planning directive. Although no quotas were set, most counties managed to select sites with good wind resources through extensive public consultation with local residents. More than 2,600 MW of capacity were identified.

Spatial planning in Denmark is performed on three levels: national planning by central authorities, regional planning by the counties, and local planning by the municipalities. The Spatial Planning Act requires the counties to designate areas in the regional plans for new energy projects. In response to the regional plan municipalities develop local wind power plans, which prescribe areas where wind turbines can be installed, for single turbines, clusters or large wind power plants, as well as conditions like hub height, colour etc.

Counties then issue zoning permits and installation permits to the municipal plan according to the act. Every county sets up guidelines for regional planning which prescribe the terms for wind turbine installations in the county. Spatial planning also emphasizes the importance of involving grid operators in order that they can prepare for the expansion by strengthening the grid so that the new turbines can be connected to the power system (see Box 4.3).

Swedish planning targets

In Sweden, the government proposed a target for electric power production from renewable energy sources (wind power, bio-fuelled combined heat and power (CHP), small hydro power) in 2002. The target was set at 10 TWh by 2010, and a planning target for wind power of a further 10 TWh by 2015 (Prop., 2002). These proposals were adopted by the parliament in June 2002. The planning target was later raised to 30 TWh, of which 10 TWh offshore.

There is no specific target for wind power development in Sweden, or how large a share of the electric power should be produced by wind turbines. The planning target just means that the municipalities should designate areas for wind power development in municipal comprehensive plans or similar documents to make it possible to produce 30 TWh a year. If this will be realized depends on what economic conditions and other rules politicians apply to wind power development.

In Sweden the municipalities have a *planning monopoly*. This means that it is the municipalities (not the counties, or the central government) that have the exclusive right to decide how land and water areas within their borders shall be used. Guidelines for land use have to be published in *municipal comprehensive plans* (MCP), which all municipalities are obliged to make. The municipalities have however to consider different kinds of public interests, and the county administration has to check that this is done. One kind of public interest is so-called areas of national interest,

Box 4.3 Municipal planning in Thisted municipality in Denmark

Thisted municipality at the Limfjord on Jutland in Denmark, has very good wind resources. When work with the municipal plan started, there were already 99 turbines on-line, most of them less than 100 kW. A work group was formed, with representatives from local interest groups and organizations, such as the nature protection association, the farmers' association, hunters, local folklore society, the wind power association, utilities, as well as civil servants from the municipality and the county.

The starting point was the map of the municipality, and the wind resource map. First zones exempted in the regional wind power plan of the county were excluded – a large zone along the sea coast, as well as minimum distances to buildings etc. Then all areas that were considered possible for wind turbines were drawn onto the map. All these sites were examined and unsuitable areas, due to local obstacles etc., were deleted.

The work group then sat around a table with the map, and discussed and negotiated until a consensus was reached on all areas that would be included in the plan. This plan was then forwarded to the building committee, which examined the proposed areas, and made some slight changes. Then the plan was exhibited to the public, before the politicians in the municipality adopted it.

specified in the Swedish Environment Law. The Cultural Heritage Law – with rules and regulations to protect and take care of buildings and areas of special interest from this point of view – can also have an impact on the planning and permission process for wind power.

Areas of National Interest

In the Swedish Environment Law, areas of national interest have been defined and specified on maps. These areas can be of national interest for nature protection, recreation, military, etc., and at sea for fisheries, shipping, etc. According to this law, land and water areas that are especially suitable for certain types of plants, for example wind power plants, shall be protected from measures (i.e. exploitation) that can conflict with the operation of such plants. This means that the construction of high buildings or factories is not permitted in these areas. Areas of national interest should be protected so that they can be used for the specified purpose.

When these laws and regulations about areas of national interest were introduced, wind power had not entered the arena, so no areas for wind power were defined at that time. However, in 2004 the Energy Agency in

Box 4.4 Criteria for areas of national interest for wind power

On land

- an average wind speed of more than 7.2 m/s 100 metres above ground
- area larger than 5 km^2 (excepting southern Sweden, which is densely populated)
- distance between wind turbines and buildings (houses and churches) shall be more than 800 metres.

Offshore

- an average wind speed of more than 8.0 m/s at 100 metres above sea level
- area larger than 15 km^2
- water depth less than 35 metres.

cooperation with the county administrations added areas of national interest for wind power to the map. In 2013 these areas were updated and based on a new wind resource map. The main criteria was that the average wind speed had to be at least 7.2 m/s 100 metres above ground level, and with exception for national parks and other protected areas (see Box 4.4).

To serve as a driver for wind power development planning on the national level, areas where it will be suitable to develop wind power should be identified and designated. On land this kind of planning is hard to conduct on a national level, since there are many local circumstances to consider, but for offshore planning on a national level it is, however, suitable. This has been done in Denmark, Germany and in the United Kingdom.

Regional and municipal planning

Many wind power projects have been developed in regions and municipalities, where there are no plans to direct where they should be sited. In Denmark several hundreds of wind turbines were installed in the 1980s, without any plans. Spatial plans were made first in 1992–4, at the request of the government. In Sweden, Germany and most other countries development has followed the same pattern. The demand for planning appears first when development has reached a level where politicians consider it necessary to direct wind turbines to the most suitable sites.

From the developer's point of view, spatial wind power plans have both advantages and drawbacks. Without a plan that expresses a political intention about the suitability of different geographical areas, the outcome of each application is uncertain. The risk of spending time and money on projects that will be denied permission, and will never generate any income, is higher when there are no plans. On the other hand it takes time, two years or more, for a municipality or region to actually draw up a plan. During this period, from when the decision to make a plan is made until it has been worked out and politically approved, it is usually impossible to get any applications processed at all, since these decisions will depend on the outcome of the planning procedure. In Denmark, development slowed down considerably during 1992–4, but after that it has been progressing more quickly, so in the long run local wind power plans should be advantageous also to developers.

Municipalities that decide not to draw up a wind power plan lose the opportunity to direct wind turbines to areas that, from a political point of view, are most suitable for this purpose. To devise a usable wind power plan, the officials involved need knowledge and should understand what criteria to use when areas are chosen for this purpose, otherwise there is an obvious risk that areas with bad wind resources and high roughness, or areas without suitable grid connection, are designated, areas that never will be exploited.

There are many planning tools available that can be used to develop wind power plans. The most important input for such plans are wind resource maps. Such maps can be developed with proper software, if wind data are available for the area. In many regions and countries the authorities have developed such maps. Relevant geographical data for use in GIS software are also necessary. However, when preparing such maps for wind power planning it is always necessary, and important, to compare GIS data and maps with reality by making surveys in the field.

Repowering plans

In Denmark and Germany, where thousands of wind turbines are on-line and some are approaching the end of their technical lifetime, planning for a second generation of wind turbines has started; this is called *repowering*. In these countries land suitable for wind power plants with reasonable wind resources have become scarce. Turbines installed in the 1980s or the beginnings of the 1990s occupy the best sites. By replacing some of these small, old turbines with fewer large ones, the number of turbines can be reduced while the power produced would significantly increase.

The technical lifetime of a wind turbine is estimated at 20–25 years. However, removing and replacing turbines before they have reached this stage has proved difficult. For the owner of the old turbines, there is no obvious reason to take them out of service. The investment has been repaid, so there is no capital cost left to pay. These turbines have become very profitable. The land lease contract for a good site is also a valuable asset.

In Denmark, the first repowering plan was implemented in 2001–2003. Owners of wind turbines with less than 100 kW nominal power received a subsidy of 2.3 Danish öre/kWh for five years if they replaced them with wind turbines with at least three times more nominal power. Owners of wind turbines with 100–150 kW nominal power received a similar subsidy

Table 4.11 Näsudden repowering phase 2

	Before	*After*
Year built	1993–96	2011
Manufacturer/number	27 WindWorld/Vestas	12 Vestas
Nominal power/wind turbine	500/600 kW	3,000 kW
Rotor diameter	37/39/42 m	90 m
Hub height	40 m	80 m
Annual power production	~30,000 MWh	~90,000 MWh

if they replaced them with wind turbines with at least twice the nominal power. The result was that 1,480 wind turbines were replaced by 272 new larger wind turbines, which increased the installed capacity from 122 MW to 331 MW.

A generational shift is also going on at Näsudden on Gotland in Sweden. It has been divided into four phases, and three have already been implemented. Before the repowering started, there were 81 wind turbines on-line at Näsudden, from small 150 kW turbines up to the huge 3 MW prototype Näsudden II operated by the state-owned power company Vattenfall. Now 59 wind turbines with nominal powers up to 600 kW have been replaced by 27 larger wind turbines with nominal powers of 1.8 MW or 3 MW. In the second phase 27 wind turbines were replaced by 12 large 3 MW turbines (see Table 4.11).

Before the repowering started, the annual power production at Näsudden was around 51 GWh/year. When all the four phases have been implemented, annual power production will increase to around 204 GWh. This generational shift has been conducted in close cooperation between a project developer, who also organized all the owners of the old wind turbines, who now are part owners of the new ones, and the planning division of Gotland municipality.

Planning methods

The development of wind power planning has followed a similar path in most places. After a period without planning, the need for planning has grown as the number of turbines has increased. This seems quite logical. An important distinction is, however, the difference between negative planning, defining areas where wind turbines should be excluded, and positive planning, designating areas suitable for wind turbines. Negative planning acts as a barrier, positive planning as a driver. Which approach that is applied depends on political decisions.

Wind turbines are rather new structures to fit into a spatial planning context. In existing regional and local plans, most areas have already been defined, and in some cases also protected, for a specific purpose; nature protection, infrastructure, agricultural land, etc. Then criteria of noise, shadow impact and minimum distances are added. When all these zones have been excluded, the areas that remain can be used for wind power. In Germany this turned out to be less than one per cent. This is, in most countries, quite a considerable area to use for designated areas for wind turbines.

The borders of areas for nature protection etc. are in many cases set quite generously and with buffer zones in addition. With experience, the actual impacts from wind turbines on different aspects of environment and impacts on neighbours have been learned, and the criteria should be adapted. Furthermore, all protected zones have a purpose, but the question that should be asked in the planning process is whether wind turbines sited in or close to such an area will have a negative impact on that *purpose*.

The most important aspect when planning for wind power is the energy content in the wind. Wind resource maps should be the starting point for all wind power planning. Small changes in average wind speed make a big change in energy content and have an even larger impact on the economic feasibility of wind power projects. There are wind resource maps available for most countries nowadays, and they have to be included in the planning methods and procedures that are applied.

A very efficient way of planning is the 'round table' method. It has been described above in Box 4.3 about Thisted municipality in Denmark. The precondition for this is that wind turbines will be built; the question is not if, but where and on what conditions. In Thisted the group worked out a plan in three meetings. The 'One Stop Shop' permission process in the United Kingdom for offshore developments can also be described as a 'round table' method.

Opinion: support and opposition

When wind power is developed it is important that people living in the vicinity of the wind turbines accept these new elements in their environment; that wind power has a high acceptance. In many countries it is easy to get the impression that there is strong opposition to wind power among the public when you read reports and letters in newspapers about protests against planned wind power projects.

If a systematic review of the press coverage is made, you will most likely find as many articles that are positive or at least neutral to the subject of wind power, as well as letters to the press that promote the development of this renewable energy source. In several countries there are some organizations that very actively oppose wind power, like Country Guardians (www.countryguardians.net) and National Opposition to Wind Farms (www.nowind.org.uk) in England, the Association for Protection of the Landscape in Sweden, and Windkraftgegner (www.windkraftgegner.de) in Germany, but they only represent a minority in these countries.

How wind turbines are perceived is a subjective matter. Different people perceive wind turbines in different ways. People can also change their opinion in time when they get used to the wind turbines. There are, of course, also wind power projects that have been planned without regard for the natural environment, visual intrusion or impact on neighbours. Sometimes people who have a positive attitude to wind power can have good reasons to oppose a specific wind power project.

Although the attitude to wind power is a matter of personal opinion and a subjective matter, it is still possible to find objective measures of public opinion by surveys carried out with proper scientific methods. *Almost all opinion polls and surveys that have been conducted so far show that a vast majority of the respondents have a positive attitude to wind power.* This applies to national surveys as well as surveys among the people in areas with many wind turbines installed in the vicinity (see Box 4.5).

According to all these surveys, most of them conducted in countries with significant amounts of wind power installed, there is a very broad support among the public of wind power in general, and awareness of the benefits of this renewable energy source. It seems that general support of wind power is high in all countries. This does not necessarily imply that the respondents would accept wind turbines in their local environment.

A Danish report (Danish Wind Turbine Owners Association, 2014) has shown that people who have wind turbines within sight of their house, school or place of work have a more positive attitude than other people. Those who live in an area with wind turbines are more positive to wind power than those who visit it or have holiday cottages there. Finally this positive attitude is stronger where people are offered the opportunity to buy shares in the wind turbines and when they have been informed about the advantages for the environment.

Many opinion polls and surveys have been carried out in areas where there are many wind turbines installed, and where people have practical experiences of wind power (see Box 4.6).

Stability of opinion over time

In a region where wind power has been developed rapidly, one would expect that acceptance among the population would change over time as the number of turbines increase. In some regions of Spain such rapid development started at the end of the 1990s, and in 2004 Spain had the fastest growth of new wind power capacity in the world. Social acceptance of wind power has been studied in three regions in Spain where many wind turbines are installed: Navarre, Tarragona and Albacete.

In Tarragona four studies were conducted from 2001 to 2003, with 600 persons in each poll. These four polls show that the strongest support comes from people living near wind power plants. In Albacete a study was

Box 4.5 Results from opinion polls in selected countries

Denmark

A nationwide survey in 2001 posed the question: 'Should Denmark continue to build wind turbines to increase wind power's share of electricity production?'

Sixty-eight per cent of the respondents answered yes, while 18 per cent found the current level satisfactory, 7 per cent were of the opinion that there were already too many and 7 per cent were undecided (Danish Wind Turbine Owners' Association, 2002).

Germany

A nationwide survey conducted in 2002 showed that 88 per cent of the respondents supported construction of more wind farms in Germany, as long as certain planning criteria were met. Only 9.5 per cent considered that there was already enough (Wind Directions, 2003).

United Kingdom

Many opinion polls have been conducted in the United Kingdom by various organizations since the first wind turbines were installed in 1991. The British Wind Energy Association has made a summary of 42 different surveys carried out from 1990 to 2002. The summary shows that 77 per cent of the public are in favour of wind energy, while 9 per cent are opposed. A survey of 2,600 persons in 2003 came to a similar result; 74 per cent of the respondents supported the government's aims to generate 20 per cent of the UK's electricity from renewable energy sources by 2020, and further development of wind power, 7 per cent were against and 15 per cent were neutral (Wind Directions, 2003).

France

In 2003 a survey of 2,090 people was conducted in France; 92 per cent of the respondents were in favour of further development of wind energy, considering both the environmental and economic advantages of the technology, but also as a substitute for other energy sources, including nuclear power (Wind Directions, 2003).

continued ...

Box 4.3 continued

United States

According to a national survey in 2005, 87 per cent of respondents considered it a good idea to build more wind farms (Yale University, 2005).

Australia

In a nationwide survey in 2003, building new wind farms to meet Australia's rapidly increasing demand for electricity was supported by 95 per cent of the respondents.

Australia has large coal mines and many coal-fired power plants and a strong lobby that protects the coal industry. However, for 71 per cent of the respondents, reducing greenhouse pollution outweighed protecting industries that rely on reserves of fossil fuels (Wind Directions, 2003).

Sweden

The attitude of Swedes is reported in reports from the SOM Institute at Gothenburg University (Hedberg, 2013). This nationwide survey that was conducted in the year 2012 showed the respondents' attitudes to different energy sources (see Table 4.12)

Of the respondents 66 per cent have the opinion that Sweden should back wind power more that today, 25 per cent that we should back as much as today and if these groups are added together, 91 per cent think that Sweden should back wind power more or as much as today.

The support for wind power in Sweden has according to these annual surveys been very high for many years (see Table 4.13).

Table 4.12 Attitudes to wind power and other energy sources in Sweden, 2012
Question: How much should Sweden back these different energy sources during the coming 5–10 years? (share in per cent, 2012)

Energy source	*Back more*	*Back as today*	*More + today (col 1+ col 2)*
Hydropower	43	51	94
Wind power	66	25	91
Solar power (PV)	81	17	98
Nuclear power	14	38	52
Biomass	42	47	89
Fossil gas	20	47	67
Oil	2	26	28

Source: Hedberg, 2013

Table 4.13 Share that want to back wind power more, in per cent

	1999	*2000*	*2001*	*2002*	*2003*	*2004*	*2005*
Back more	74	72	71	68	64	73	72
	2006	*2007*	*2008*	*2009*	*2010*	*2011*	*2012*
Back more	77	79	80	74	66	70	66

Source: Hedberg, 2013

Box 4.6 Opinion on wind power, selected areas

United Kingdom

In the United Kingdom fewer than two out of ten people would oppose development of wind power in the vicinity of their house. More than a quarter would have a very positive attitude to that according to a national survey from 2003. Respondents who lived in areas where there already were wind turbines installed, 94 per cent would be positive to a further development and only 2 per cent were negative (Taylor Nelson Sofres, 2003).

Scotland

According to a survey conducted for the Scottish Executive, 82 per cent of those who lived close to Scotland's ten largest wind power plants wanted more electricity generated by wind power, and 50 per cent supported an increase of the number of turbines at their local wind power plant.

The poll covered 1,800 persons, living within three zones, up to 5 km, 5–10 km and 10–20 km away from operating wind farms. People who lived there before the wind power plant was developed say that although, in advance, they thought that problems might be caused by its impact on the landscape (27 per cent), traffic during construction (19 per cent) and noise during construction (15 per cent), in reality these figures after construction were 12 per cent, 6 per cent and 4 per cent respectively (Wind Directions, 2003).

Aude region, France

In a sample of 300 persons living near wind turbines in the Aude region in southern France, 46 per cent agreed that wind turbines affect the countryside, and 55 per cent considered wind farms to be aesthetically pleasing (Wind Directions, 2003).

Table 4.14 Public acceptance of wind turbines in Navarre, change over time

	1995	*1996*	*1998*	*2001*
Turbines*	6	72	187	659
Positive %	85	81	81	85
Negative %	1	2	3	1
Indifferent/ don't know %	14	17	16	14

*Mostly 660 kW turbines

conducted in 2002. It showed that 79 per cent of the respondents considered wind energy to be beneficial (EWEA, 2009).

In Navarre a 2001 study showed that 85 per cent were in favour of the implementation of wind power in Navarre, and 1 per cent were against. The study also showed that acceptance increases while new wind power plants are developed and installed. For most people the benefits of wind energy compensated for any negative impacts experienced during implementation (see Table 4.14).

From 1995 to 2001 the number of turbines increased from 6 to 659, while the share of the inhabitants that have a positive attitude to wind power has remained constant (EWEA, 2009).

Wind turbines in the living environment

In 2004 Gotland University made case studies in three different areas on the island of Gotland in the Baltic Sea, where people were living close to wind turbines. In the village of När, all who lived within 1,100 metres of two large wind turbines were interviewed, in Klintehamn a sample of those who were affected by shadow flicker at sunset, and in Näsudden those situated in the middle of a large wind power plant with 81 turbines. In total 94 persons in 69 households were included in the study (Widing et al., 2005).

Considering that all respondents live close to wind turbines, the nuisances reported were surprisingly small. Very few of them were annoyed by noise, shadows or consider that their view of the surrounding landscape had been destroyed. Of the total number of persons interviewed, 85 per cent were *not* annoyed by noise from the wind turbines around their homes. For shadows flicker the share of *not* annoyed was even bigger: 94 per cent. Quite a few of the persons living at Näsudden, where there were 81 wind turbines on-line, 13 per cent think that their view of the surrounding landscape has been negatively affected. Of all persons in all three areas, 89 per cent expressed the opinion that wind turbines had not spoiled their view. The acceptance of wind power among people living as close neighbours to the wind turbines was high (Widing et al., 2005).

Wind power and tourism

Since wind turbines should be installed at sites with good wind conditions, many wind farms are sited at the coast, onshore and in recent years also offshore. Many coastal areas are also popular tourist resorts, so there is an obvious risk for a conflict of interest. The same conflicts can occur in mountain areas, in Scotland and in skiing resorts. Several surveys have been carried out to find if wind turbines will deter tourists from attractive holiday areas.

A survey among tourists in Germany in 2003 showed that 76 per cent considered that nuclear and coal-fired power plants spoiled the landscape, whilst only 27 per cent thought the same about wind turbines (Wind Directions, 2003).

A survey on tourism in Schleswig-Holstein showed that the wind industry does not affect tourism in the region. Visitors are aware of the increasing number of turbines, but it does not influence their behaviour (EWEA, 2003).

In Belgium there were plans to develop an offshore wind farm 6 km off the coast where there are many holiday resorts (this wind farm has now been built). A survey conducted in 2002 by the West Flemish Economic Study Office before it was built showed that 78 per cent of the public were either very positive or neutral to this offshore wind farm (see Table 4.15).

Two separate polls have looked at the impacts on tourism in Scotland. A MORI poll in 2002 found that over 90 per cent of visitors would return to Scotland for a holiday whether or not there were wind power plants in the area. Another survey by Visit Scotland tourism agency found that 75 per cent of visitors were either positive or neutral towards wind power development in general, but less positive to their visual impact. However, those who actually had seen wind turbines during their visit were more positive than those who had not (EWEA, 2003).

There are other surveys that have investigated how different groups of inhabitants evaluate the view of the landscape. They show that permanent residents in the countryside consider the landscape as a natural resource that

Table 4.15 Public perception of Belgians of wind farm 6 km off the coast

Group	*Negative*	*Neutral to positive*
Residents	31.3	66.5
Second residence	10.2	88.8
Frequent tourists	18.7	81.3
Occasional tourists	19.5	80.5
Hotels etc. with sea view	19.5	80.5
Other	15.3	84.7
Total	20.7	79.3

Source: EWEA, 2003

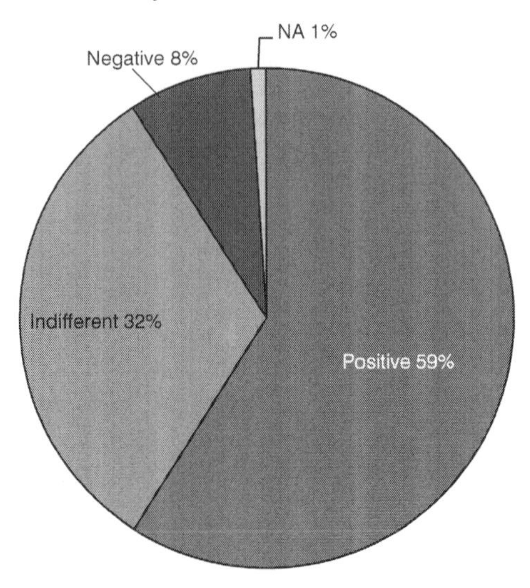

Figure 4.8 Perceptions of wind power on Gotland. On Gotland 59 per cent of tourists considered that wind power in the landscape gave a positive impression, for 32 per cent it didn't matter and 8 per cent considered wind turbines to have a negative impact. One per cent had no opinion

should be utilized in a sensible way, while those who use the landscape for recreation have a more aesthetic view and consider it as a 'picture postcard' that should remain unchanged. According to the surveys made in Scotland and Schleswig-Holstein, it seems that wind turbines can be an accepted element in the tourists' picture postcards (Hammarlund, 2002).

On Gotland a survey among tourists was made in 2013 by Vendula Braunova from Uppsala University. Throughout the month of July questionnaires were given to tourists who were leaving the island on the ferry. Of the 735 tourists who responded to the questionnaire 83 per cent had seen wind turbines during their stay on the island. The first question was how they perceived wind turbines in the landscape. It turned out that 59 per cent said that they made a positive impression, 32 per cent that it didn't matter, while 8 per cent considered wind turbines to have negative impact on the landscape (see Figure 4.8) (Braunová 2013).

Another question was whether the wind turbines would influence their decision to visit Gotland again, that is, whether wind power has a negative impact on tourism. The result showed that 59 per cent didn't care, 19 per cent thought that wind power was a reason for them to *revisit* the island, and only 2 per cent claimed that they did not want to return due to the wind turbines. The conclusion was that wind power has a *positive* impact on tourism.

Box 4.7 NIMBY – 'not in my back yard'

- NIMBY A: Positive attitude to wind power installations in general, but negative attitude to installations in the close vicinity.
- NIMBY B: Generally negative attitude to wind power.
- NIMBY C: A positive attitude to plans to develop wind power, that changes to negative when there are plans to install wind turbines in the vicinity.
- NIMBY D: Negative attitude to the planning procedure rather than to wind power.

NIMBY – 'not in my back yard'

Persons who have a positive attitude to wind power in general sometimes can have another attitude to development of wind power close to their house or holiday cottage. This phenomenon is known as NIMBY – 'not in my back yard'. The Dutch researcher Maarten Wolsink (2000), who has studied the NIMBY phenomenon, defines four different types of NIMBY reactions (see Box 4.7).

This NIMBY effect can be measured. A typical NIMBY curve starts at a rather high level. In a survey on the general attitudes to wind power within a group of people, say 75 per cent express a positive attitude. When a project to install wind turbines in the vicinity starts, some of these people get worried about the impact on their living environment: noise, shadows, effects on their view etc. The proportion of people with a positive attitude tends to decrease to around 60 per cent. When the wind turbines have been installed and have been running for a couple of months, the proportion of people with a positive attitude recovers to the initial value and often actually goes even higher than that.

The general public has a very positive attitude to wind power according to surveys in many different countries, regions and local areas. If those that are concerned are informed about the environmental advantages – that wind turbines produce electric power without any hazardous emissions – attitudes tend to be even more positive. Attitudes also depend on how people in the vicinity of a planned wind power project are informed about the plans.

The problem of NIMBY has been under discussion for some time. There is no doubt that a NIMBY reaction can occur among residents in areas where new wind power developments are planned. However, if wind power development in a country is slow and the targets set are not met, this cannot be blamed on resistance from local communities. Rather, the reasons are institutional barriers: government policy; the laws and rules that regulate the permission process; and the economic conditions for wind power.

Institutional factors have greater impact on wind energy siting than public support, according to the researcher Maarten Wolsink (Wolsink, 2000).

The federal state of Schleswig-Holstein in northern Germany had 1,800 MW wind power installed in 2002, which produced 30 per cent of the state's energy consumption. A study (Eggersglüss, 2002) showed that most people accept the siting of wind turbines if the following principles are followed:

- sufficient distance to residential areas;
- quiet turbines are chosen;
- the population is kept properly informed;
- there is a financial benefit for the community;
- the developer is from the area;
- landowners' views are sought when site is chosen.

If all wind power project developers followed these simple guidelines, acceptance at the local level would probably be as high as the general acceptance of wind power.

Acceptance problems at higher levels

Surveys that have been conducted to find out the public opinion on wind power prove that acceptance of wind power by the general public is not a problem. Wind power has in fact a very strong *support* from the public. Problems with acceptance are usually found at higher levels, although this obviously differs in different countries.

In Sweden and many other countries as well, authorities, politicians, grid operators, power companies and industry have problems in accepting wind power and they raise barriers that will delay or stop its development. Also among scientists working in the energy field acceptance is low, as many scientists have strong interests in conventional power technology, such as fossil fuel combustion and nuclear power. Many leading scientists are still repeating the same unfounded arguments against wind power that power companies have used since development started in the 1970s. However, wind power has now become a mainstream technology and even big transnational power companies have made large investments in wind power plants.

There is also a conflict of opinion between those who work for nature protection and those who work for environmental protection. Many biologists and nature protectionists still think that nature can be protected by 'fencing in' and don't seem to understand these fenced in-areas will be affected by climate change (i.e. changes in temperature, humidity, precipitation, etc.), which will change the living and survival conditions of local flora and fauna.

References

Braunová, V. (2013) Impact study of Wind Power on Tourism on Gotland. Master's thesis, Uppsala University.

Danish Wind Turbine Owners' Association (2002) 'Hvem ejer vindmöllerne?' Accessed 7 December 2014 at http://www.dkvind.dk/fakta/07.pdf

Danish Wind Turbine Owners' Association (2011) 'Vindmöller og drivhuseffekten'. Accessed 7 December 2014 at http://www.dkvind.dk/fakta/M2.pdf

Danish Wind Turbine Owners' Association (2013) 'Danskernes mening om vindmöller'. Accessed 7 December 2014 at http://www.dkvind.dk/fakta/M6.pdf

Eggersglüss, W. (2002) 'Das steht und das dreht sich' in Ministerium für Finanzen und Energie des Landes Schleswig-Holstein (ed.) *Stimmen zur Windenergie*, pp. 8–10.

European Commission (2013) Communication from the Commission. Delivering the internal electricity market: making the most of public intervention. Draft 1. Brussels: European Commission.

EWEA (2003) 'Public acceptance in the EU' in *Wind Energy – The Facts – Environment*. Brussels: EWEA.

EWEA (2005) *Support Schemes for Renewable Energy*. Brussels: EWEA. Accessed 22 December 2014 at http://www.ewea.org/fileadmin/ewea_documents/documents/projects/rexpansion/050531_Rex-final_report.pdf

EWEA (2009) *Wind Energy – The Facts*. Brussels: EWEA.

ExternE (2012) *Externalities of Energy*. Luxembourg: European Environment Agency.

Freund, H.D. (2002) 'Einflüsse der Lufttrübung, der Sonnenausdehnung och der Flügelform auf dem Schattenwurf von Windenergieanlagen'. *DEWI* no. 20.

Gammelin, Cerstin (2013) 'Oettinger schönt Subventionsbericht'. *Süddeutsche Zeitung*, 14 October. Accessed 7 December 2014 at http://www.sueddeutsche.de/wirtschaft/foerderung-der-energiebranche-oettinger-schoent-subventionsbericht-1.1793957.

Greenpeace (2005) *Wind Force 12*. Brussels: EWEA Publications.

Hammarlund, K. (2002) 'Society and wind power in Sweden', in Pasqualetti, M., Gipe, P. and Righter, R. (eds) *Windpower in View*. London: Academic Press.

Hedberg, P. (2013) 'Fortsatt stöd för mer vindkraft', in Weibull, L. Oscarsson, H, and Bergström, A. (eds) *Vägskäl*. Göteborg: SOM-institutet, Göteborgs universitet.

Holttinen, H. (2004) *The Impact of Large Scale Wind Power Production on the Nordic Electricity System*. Helsinki: VTT Publications.

Länsstyrelserna i Skåne (1996) *Lokalisering av vindkraftverk och radiomaster i Skåne*. Lund: Länsstyrelserna i Skåne.

Miyamoto, C. (2000) *Possibility of Wind Power: Comparison of Sweden and Denmark*,.Lund: IIIEE, Lund University.

NWCC (National Wind Coordinating Committee) (2001) *Avian Collision with Wind Turbines: A Summary of Existing Studies and Comparisons to Other Sources of Avian Collision Mortality in the United States*. Washington, DC: West Inc.

Pasqualetti, M., Gipe, P. and Righter, R. (2002) *Wind Power in View*. London: Academic Press.

Pettersson, J. (2005) *Waterfowl and Offshore Windfarms: A Study 1999–2003 in Kalmar Sound, Sweden*. Lund: Lund University.

Remmers, H. and Betke, K. (1998) 'Messung und Bewertung von tieffrequentem Schall', *Fortschritte der Akustik – DAGA 98*. Oldenburg: Deutsche Gesellschaft für Akustik.

Rönnborg, P. (2009) *Det där ordnar marknaden*... Göteborg: Gothenburg University.

Sagrillo, M. (2003) *Putting Wind Power's Effect on Birds in Perspective*. AWEA.

SOU (1999) *Rätt plats för vindkraften*. Stockholm: Fakta info direkt SOU.

Taylor Nelson Sofres (2003) *Attitudes and Knowledge of Renewable Energy amongst the General Public*. London: TNS.

Widing, A., Britse, G., and Wizelius, T. (2005) *Vindkraftens miljöpåverkan: Fallstudie av vindkraftverk i boendemiljö*. Visby: CVI.

Wind Directions (2003) 'A summary of opinion surveys on wind power', September/October, 16–13.

Wizelius, T. (2014) *Wind Power Ownership in Sweden: Business Models and Motives*. London: Routledge.

Wizelius, T., Britse, G., and Widing, A. (2005) *Vindkraftens miljöpåverkan – utvärdering av regelverk och bedömningsmetoder*. Visby: CVI.

Wolsink, M. (2000) 'Wind power and the NIMBY myth: institutional capacity and the limited significance of public support'. *Renewable Energy* 21(1): 49–64.

WWF (2006) *Bird Species and Climate Change*. Fairlight, NSW: Climate Risk.

Yale University (2005). *Survey on American Attitudes on the Environment: Key Findings*. New Haven, CT: Yale University School of Forestry and Environmental Studies.

5 Wind power in the electric power system

Wind turbines are not only a new kind of power plant that transform wind energy into electric power, they also have some other characteristics that power companies, utilities and grid operators are unused to. The wind speed changes all the time in a way that is hard to predict, so the power production will vary. Wind turbines are comparatively small and are often connected to the distribution grid, while large conventional power plants are connected to the transmission grid, with much higher voltage levels.

The electric power system

The electrical power system can be described as a socio-technical system, divided into three interrelated subsystems:

1. technical system
2. legal/political system
3. economic system.

The technical system works the same way in all electrical power systems throughout the world, as it is governed by the laws of nature and common standards are applied, although grid frequency, voltage levels and other technical parameters may differ. The legal/political systems and the economic systems vary in different countries.

These three subsystems interact with each other (see Figure 5.1).

The technical system consists of three different components:

1. power plants which produce the electric power;
2. power grid, which transmits power to consumers;
3. electrical equipment, used by power consumers.

Electricity is not an energy source, but is utilized to transport energy in the form of electric power from power plants to power consumers. The electric power grid constitutes a practical and cost-efficient means

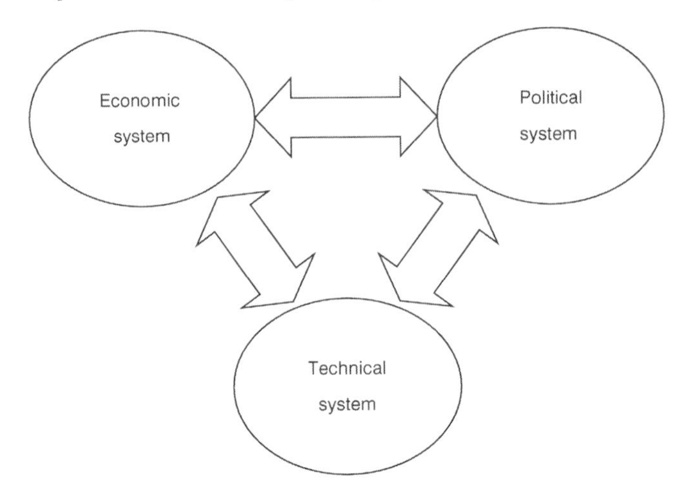

Figure 5.1 Socio-technical system

to transport power from plants to consumers, who transform the electric power into light, heat or mechanical work with different types of equipment.

When the use of electric power started over a hundred years ago, the power grid was built only in close vicinity to power plants. The grid at that time consisted of small isolated 'islands' in cities and around hydropower stations where there were factories, which had earlier exploited the energy in the running water by water wheels to run sawmills, blacksmiths' workshops, etc.

Such local grids were very vulnerable. If the river ran dry or the power plant had a failure, or needed service, no electric power could be produced. To avoid this, back-up power plants were needed. Some decades later these isolated grid-islands were connected to each other to form larger regional and finally national power grids. The need for back-up decreased and was almost eliminated, as the many power plants connected to the grid could replace each other. Today the power grids are interconnected across national borders.

The electric power system is, from a technical point of view, complicated and complex. The development of this system has contributed immensely to the development of the modern industrial society and to the social wellbeing of mankind. The impacts on social development are immediate and important; a single electrical light bulb in a rural village in a developing country will make it possible for children to do their homework and to become educated.

In the developed world most people take electricity for granted, and don't realize its role until there is a blackout. Those who operate the electric power system carry a large responsibility, and it is understandable that electric power engineers are careful and rather conservative when it comes to the introduction to this system of new technology, like wind turbines, with some unknown or unusual properties.

In Sweden, and most other countries, the backbone in the power system consists of a *national transmission grid*, with a voltage level of 400 or 230 kV that is managed by a transmission system operator (TSO). In Sweden this role is fulfilled by the state-owned public utility Svenska Kraftnät (Swedish National Grid). The very high voltage transmission grid is used to transmit large amounts of power over long distances.

The next level is the *regional grid*, with a voltage level of 130 or 70 kV (in Sweden) which transmits power from the national transmission grid to the *local distribution grids*. These have lower voltage levels, 40, 20 or 10 kV, and distribute power to factories, households and other consumers. Before the power enters the consumers' low voltage grid, the power is transformed to 690V for industry or 400V (230V per phase) for other electric power consumers (see Figure 5.2).

To transmit large amounts of power over long distances it is most cost-efficient to use as high voltage as possible. With all transmission of electric power, losses (in the form of heat) are unavoidable. How large these losses will be depends on the dimensions and length of the power lines. In the Swedish power system around 10 per cent of the power that is fed into the grid will be lost. The TSO is actually the largest power customer in Sweden, since the power that is lost has to be bought and paid for.

In Sweden the power system is based on a few very large hydropower plants in the north of Sweden and nuclear reactors in the south. There are also some CHP plants that produce heat and electric power simultaneously and some gas turbines that are used for peak power when the loads are extremely high during cold winter days or to balance sudden increases in power consumption.

Nowadays the power grids in different countries are interconnected, and this makes it possible to import or export electric power when there is a shortage or surplus in the national power system.

Electric power consumption

Power consumption is always changing, during each day and also during the year. The need for power is, of course, less in the night-time when factories are idle and people are sleeping. In the morning, when people prepare breakfast, factories, shops and offices start to operate, power consumption increases quickly, and in the evening it decreases again, interrupted by an increase at mealtimes (see Figure 5.3).

In Sweden where many dwellings use electric heating, electric power consumption is much higher in the winter than in the summer. The peak usually occurs in January or February, when it is cold throughout the country and the wind is blowing as well. Then power demand is at its highest and all power plants run at full capacity, and some power will also be imported (see Figure 5.4).

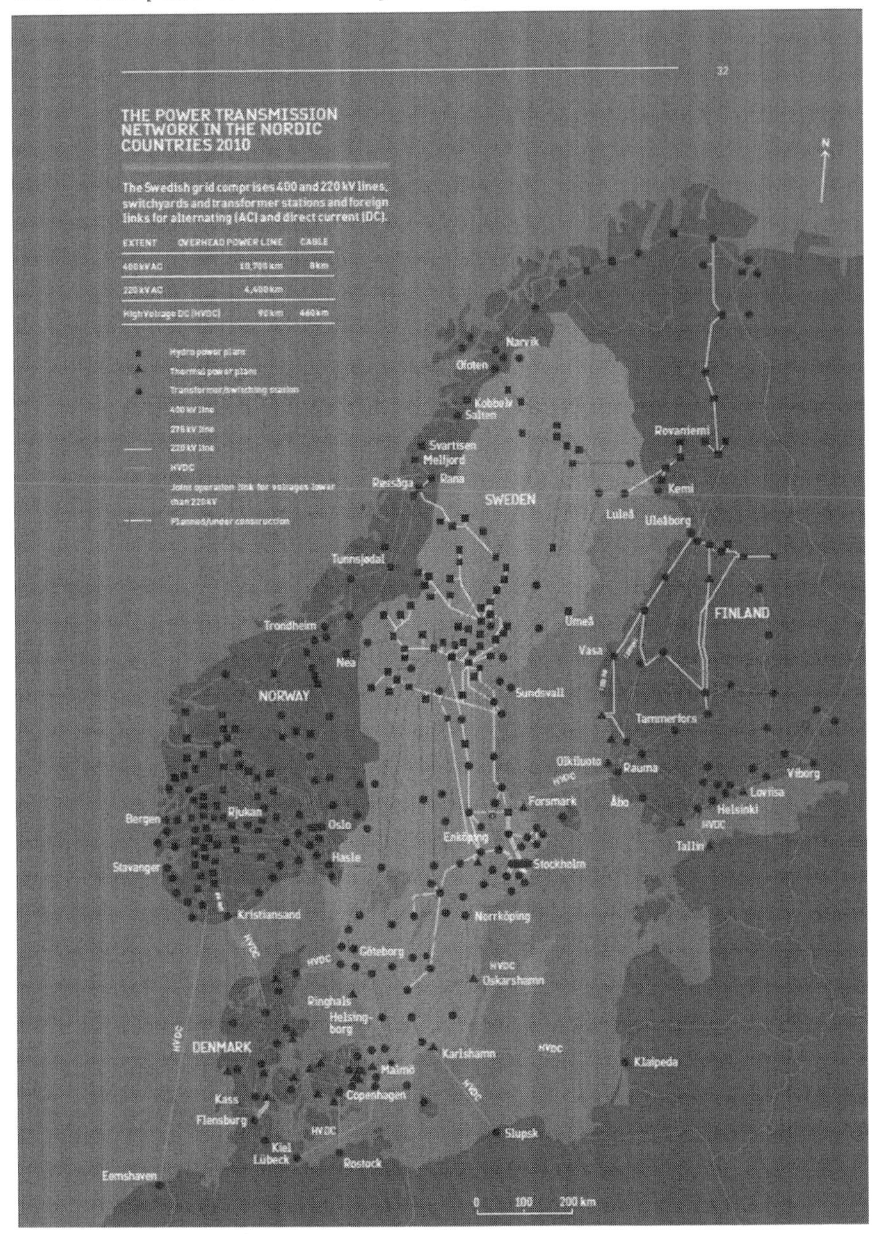

Figure 5.2 The power network in the Nordic countries 2010. The electric power systems in the Nordic countries, Sweden, Denmark, Norway and Finland, are interconnected (Nordel) and have a common power market (Nord Pool). There are also links to power systems in other countries, for import or export of electric power (illustration: Svenska Kraftnät, 2011)

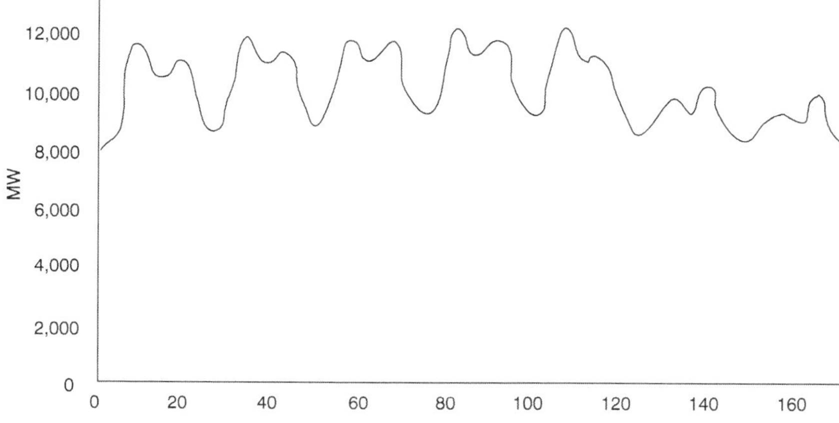

Figure 5.3 Variations of power consumption, diurnal and weekly. The diagram shows the variations of load during a normal week. The vertical axis shows the power demand (consumption) and the horizontal axis the hour of the week, starting 00.00 Monday. The peaks are significantly lower on weekends (illustration: Söder, 1997)

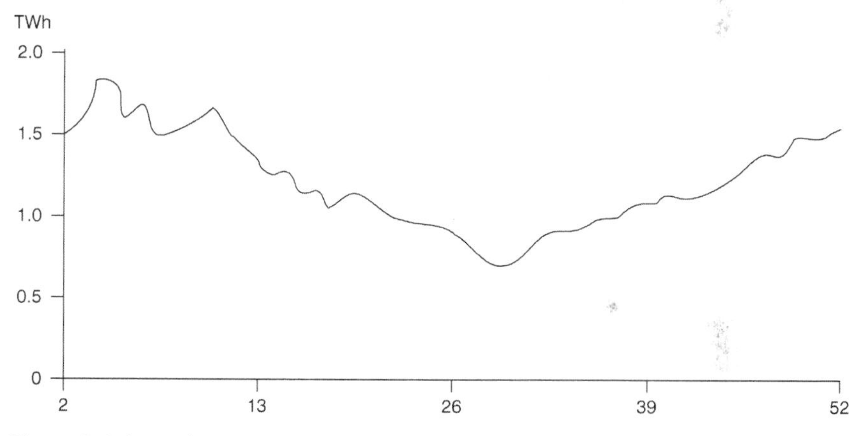

Figure 5.4 Annual variations in power consumption. The diagram shows the weekly variations of power consumption in a region in Sweden during one year. In a cold week in winter, power consumption is more than twice as high as during the summer holiday weeks (ilustration: Blomqvist, 2003).

Balancing power production and load

Electric power is a perishable commodity; power production in the grid has to match the power consumption at all times. If not, the frequency will change. If the power production is larger than the consumption, the frequency will increase; if consumption is larger than production, the frequency will decrease. In an electric power system frequency is universal, i.e. exactly the same at every place in the system. The voltage level is local, and can vary at different places (nodes) in the system. When there is a mismatch between supply and consumption (load), electric equipment can be damaged or the grid breaks down and a blackout occurs. However, this happens very seldom and the grid operators know how to manage the power system and to handle the balance of supply and demand.

Different power plants have different roles in this system: to supply *base load, power regulation* or *peak load.* In Sweden nuclear power plants are used for base load, they run day and night most of the year, and it's a slow and complicated matter to regulate the power output of a nuclear reactor. Hydropower also supplies some base load, but is also very easy to regulate, and hydro power plants are used for this purpose. Gas turbines can be started within seconds, and are used for peak power and when there is a need for very fast power regulation. Nowadays peak loads, or rather power on the margin, are often handled by imported power from neighbouring countries. It is also possible to regulate power consumption on the demand side, for example by turning down or off some large industrial loads.

To be able to follow changes in the load (consumption), power reserves have to be available. *Primary reserves* are power plants with very short start up times, from 1 second to 1 minute; these are used for fast regulation. The power plants used for this are gas turbines and hydropower. These are replaced by *secondary reserves*, power plants with longer regulation times, 10 minutes to 1 hour, and with lower operational costs. When the secondary reserves increase their production the primary reserves are replaced.

To prevent a blackout, if a failure occurs at some large power plant so that it has to close down at short notice there is always a *disturbance reserve* in the power system as well. This disturbance reserve is sized by the largest power plant that could fail. If a large power plant has a breakdown and no longer produces any power, there is always a power plant in reserve that can start to operate and replace this power. In the Nordic power system the disturbance reserve is 1,200 MW, the size of the largest power plant (a nuclear reactor) in the power system.

Wind power in the electric power system

Wind turbines are quite small power plants that should be sited where the winds are strong. Therefore they are often located at the periphery of the grid, where the grid is weak. One or a few turbines can always be connected directly to the distribution grid, but large wind power plants have to be connected to the higher voltage level on the regional grid.

The power output from a wind turbine varies with the wind speed. Wind energy can't be used for base loads, since the production varies. It can't be used as regulating power, since the energy from wind turbines cannot be increased with demand. It can't be used as peak power either, for the same reason. From the electric power system point of view, wind power works in

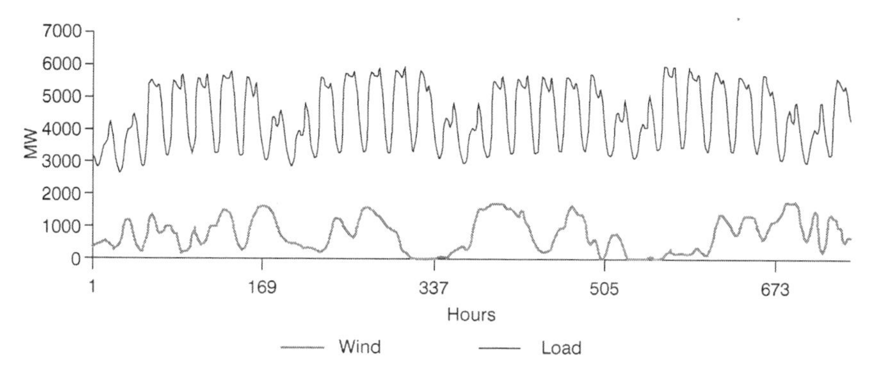

Figure 5.5 Wind power variations. This diagram shows how loads (consumption) and wind power production varied in January 2000 in Denmark. The variations are handled just like ordinary load variations. When wind turbines produce much power the contribution from other power plants is reduced, and vice versa (illustration: Holttinen, 2004)

a completely different way than conventional power plants. Wind turbines are usually connected to the distribution grid, the same power lines that consumers (households, industries) use. They function as *distributed* power plants.

The grid operator can handle the variations of power produced by wind turbines in the same way as variations in load (power consumption). It is simply treated as negative load. When the wind speed decreases, the power from other power plants, for example hydropower, will be increased. As long as the proportion of wind power is below 10 per cent of the total power production the intermittency is no problem (see Figure 5.5).

Power companies and grid operators are not used to this (at least not in countries and regions where wind power development is in its infancy). However, there is no problem in integrating quite a lot of wind power into the power system.

When the wind power penetration (wind turbines' share of the power production) reaches a level of 10 per cent, the grid and power system may need to be adapted to wind power. During the last five years countries like Denmark, Germany and Spain have crossed this threshold. In some regions wind power penetration is higher than that, like the German Länder of Schleswig-Holstein and Nordfriesland, but these regions are also integral parts of a much larger power system which can absorb this production. The surplus power is exported to other regions or countries.

There is however no problem in using a much larger proportion of wind power in the power system. Power systems have a good capacity to regulate power to keep supply and demand in balance; this is done every minute of the year, with or without wind power, and there are always power plants with reserve capacity in the power system.

Local production for local consumption

A person or a company, for example a farm or an industry, can install wind turbines on its own land, and use some or all of the power produced by the wind for its own power consumption. When the wind turbine produces more than its own requirements, the surplus is fed to the grid, and vice versa. In such cases the grid works like an energy storage facility, like a battery. Such wind turbines are connected to two meters, one that registers the power that comes in (which is bought), and another that registers the power fed to the grid (which is sold). The value of the power for own use is higher than of the power sold, since it replaces purchased power including grid fees, energy taxes and VAT added to the price.

Some grid operators accept pairing the power that is delivered to the grid with the power delivered from the grid. In such cases, the meter simply runs backwards when surplus production is fed to the grid; this is called *net metering*. In the United States 36 of the federal states have laws concerning net metering. In California a limit is set at a nominal power of 10 kW, in Arizona the upper limit is 100 kW. In other states the limits are somewhere in between, but in Iowa, New Jersey and Ohio there is no limit for maximum nominal power.

Net metering is a cheap and simple method which encourages consumers to invest in wind power and other renewables like photovoltaic (PV) cells. It is also easy to administer for grid operators and it makes sense from a system point of view to use the grid for energy storage rather than installing battery banks for this purpose.

Factories with an even and large power demand can use all power from a wind turbine for its own use as well. On Gotland there is, for example, an animal feed factory at the harbour of Klintehamn, that runs day and night all year around, that has its own wind turbine (500 kW) and feeds the power directly into the factory where all the power will be used. The factory is of course also connected to the grid; otherwise the machines would stop when the wind stops blowing. A sawmill nearby also has an own wind turbine on its premises. During the last few years some large industries have also installed wind turbines on their premises, such as the paper pulp factory in Skutskär, where five big 2 MW turbines deliver power directly into the factory.

In India a policy named *banking and wheeling* of electricity from wind power plants is used. Banking means that power generated by a private company at one time of the year can be used at another time, i.e. the utility grid can be used for 'storage'. Wheeling means the right to use the utility grid to transmit power from a wind power plant to a factory elsewhere. This has enabled so-called *captive consumption*; companies can produce their own power, and transport it through the grid to their factories.

Management of the technical system

The transmission system operator (TSO) has the overall responsibility for the operation of the electric power system. In Sweden Svenska Kraftnät is the TSO, and also operates the very high voltage transmission grid. The regional grids can be operated by different companies in different regions, like Vattenfall, Fortum, Eon. The local grid operators, owned by large power companies or municipal utilities, manage the distribution grid.

The principle for the power system is that the production with low operating costs is run all the time and the production with higher operating costs is used when demand increases, according to the so-called *merit order system*.

In a deregulated market, the task of balancing supply and demand is distributed to balance responsible players (BRP) and power trading companies. All production and consumption goes through a trader who has a contract with a BRP. These companies (BRPs) give production schedules to the system operator (TSO) one day ahead, but they can change the schedule up to the hour of delivery. During the operating hour, the responsibility is moved to the TSO.

Integration of wind power

Wind turbines are rather small, with an average nominal power of around 2 MW in 2010, compared to conventional power plants with nominal power from around 10 MW to over 1,000 MW.

Wind power is a modular technology; wind power plants can consist of a single wind turbine, small groups of turbines or large wind power plants with hundreds of turbines. Single and smaller groups of wind turbines are connected to the distribution grid, to provide local production for local consumption: *distributed generation*.

Availability differs as well. Power from wind turbines is available only when there is wind, and varies very much with the wind speed, since the energy in the wind is proportional to the *cube* of the wind speed. This means that wind power can neither be used as base power, regulating power or peak power in the electric power system. Instead the other components (power plants) have to adapt their production to the actual power production from wind turbines.

Since wind turbines convert energy in the wind to electric power, they don't need any fuel, thus no transports of fuel are needed, and costs are not dependent on the availability and price of raw materials on the world market. The wind is a local resource for local use (although it may be transmitted far in a power distribution system).

Among the renewables, wind power and solar energy have this property, while renewable fuels from biomass etc. are traded on a market and transported to power plants.

Wind power has these specific properties:

- moderate size (and price)
- modularity
- intermittency
- distributed generation
- local resource utilization
- independence of raw materials market (and price)
- no fuel transport
- no emissions to environment.

These properties define the role of wind power in the electrical power system, and in the global energy- and ecosystem.

How do these properties and their impact fit into the electric power system?

Moderate size and price

The moderate size and price of wind turbines opens up the opportunity for new actors to enter the power system, given that laws allow independent power producers to connect to the electric power grid, which in turn will add wind turbines to the power system.

Modularity

Modularity – which means that a wind power plant may consist of anything between one and 1,000 wind turbines – adds flexibility to the power system, with shorter lead times for development than, for example, nuclear power plants. It is easier to adapt power production capacity to a changing demand. To fill this role, however, the legal/political system must enable a fast process of granting permission to such projects.

Intermittency

Intermittency – power production varies with the wind – will put new demands on grid operators as well as wind turbine manufacturers to enable a high penetration of wind power in the power system. With higher penetration, wind turbines have to be able to contribute to the stability of the power system and participate in power regulation to control voltage stability and frequency. Wind turbine manufacturers have thus developed technology to accomplish such demands.

The value of wind-generated electric power can be increased with good wind forecasts that could make it possible to predict power production over coming hours or days with accuracy. Forecasts will, of course, not have any impact on the production of the turbines, but it will have a considerable economic value. If predictions and actual production coincide, the penalties for deviations from 'ordered' power will decrease.

Distributed generation

In the early days of wind power, when the turbines were quite small (20–250 kW), they were intended to produce local power that could be used in the close vicinity, and even today many turbines still function in that way. This is an obvious advantage for the power system, since it will reduce the losses in the grid (the longer distance power is transmitted, the larger are the losses). As long as the maximum power from the turbines is less than the minimum load in the local grid, all power from the wind turbines will be used in the local area.

A factory with a large and relatively constant power demand can use all power from a wind turbine for its own use. Some factories which operate day and night all year round have their own wind turbines and feed the power directly into the factory where all the power will be used.

Local resource utilization

Wind turbines utilize local wind resources. The advantage is that no raw materials have to be transported to the site during its lifetime. This saves a lot of costs for and emissions from transport vehicles, but is also an advantage for energy security.

Independence of raw materials market (and price)

Since no raw materials such as coal, oil, gas or biomass have to be used, and the wind is free, the costs of production are predictable. The capital costs and costs for operation and maintenance are known in advance and regulated in contracts. This property is also an advantage for energy security.

No fuel transport and no emissions to environment

Both these properties are advantages for the environment, and reduce emission of greenhouse gases and hazardous emissions. Since reduction of GHG emissions is on the political agenda, these properties give politicians a good motive to introduce support to wind power, which will add more wind turbines to the power system.

Technical acceptance

In 2005/06, wind power became accepted as a component in large electric power systems by most operators in the world. To achieve a high penetration of wind power, there are still some requirements that have to be met and additional equipment in the wind turbines will be necessary. The technology for large-scale integration in the power system has, however, already been developed and implemented, for example in Spain, and will sooner or later be used also in the rest of the world.

The last technical barriers to wind power have been almost removed, and for the future there are rather local conditions, like wind resources and the power mix, which will define the amount of wind power that can be integrated in electric power systems in different countries.

The electric power market

In the early days of the electric power system, there was fierce competition between different power companies; each had their own power grids, voltage levels etc. As the system grew, politicians found it necessary to bring order to this chaos and regulate the market. Electric power was turned into a *public service*, and prices were negotiated between power companies and politicians.

At that time the supply and distribution of power was a strictly national matter, regulated by the state and considered as a matter of public service and of national security. Now in Sweden and many other countries the electric power market has been deregulated and market places for electric power, like Nord Pool in Scandinavia, have been created.

Market deregulation

During the early 1990s the deregulation of the electric power markets was set on the political agenda, in the European Union, the United States, Sweden and elsewhere. Deregulation advocates argued for a radical change of policy. The first step would be to separate power production, distribution and trade. Power plants would compete and get an incentive to cut costs that would reduce prices for power. All power producers would get free access to the grid. Finally electric power should be a commodity, which was bought from power producers and sold to customers by independent power trading companies. The final step would be to open the boundaries, interconnect the power grids in different countries and regions to create an open international market for electric power.

The basic principles of a deregulated market are:

- production, distribution and sale are separated, and no company may have more than one of these roles.
- every power producer has the right to connect to the power grid, if it is technically possible.
- every consumer may choose any company to buy power from.

Grid operation is treated as a *natural monopoly*, and for each region a specific grid operator has an exclusive right to operate the power grid.

These concessions are granted by a public agency (the state). A public agency (the state) ensures that regulations are observed and that grid tariffs are reasonable.

On the electric power market, electric power is a *commodity*, with no direct relation to the actual physical product electric power (the oscillating electrons in the power grid).

In some countries, such as Sweden, Norway and the United Kingdom, and some states in the USA such as California, the markets have been completely deregulated. Other countries have just started deregulation and there are countries that resist this process.

One important question that has to be solved in the deregulation process is who will have the responsibility for the overall system balance and how can this be maintained. In a power system, supply and demand has to balance at all times, otherwise the system will collapse.

The power crisis in California after deregulation, where the power prices skyrocketed and many huge power companies, like Pacific Gas and Electric, went bankrupt, proved that deregulation is a tricky business. To avoid such situations, someone has to have the overall responsibility and the legal power to order producers to supply the power needed to balance demand.

The deregulation process will most likely continue, and will work when rules and regulations have been properly adjusted. Whether this will lead to lower power prices for end users is, however, not certain.

In a public service system, the actual cost of production will set the price; in a market system it is supply and demand. When demand increases, prices will increase independent of the cost of producing the power. When demand decreases, prices will fall. However, power companies also have the option to lower supply to keep prices on a higher level. The questions of whether the electric power market actually works, if there is a free competition, if prices are fair, has been the matter of much public debate in Sweden.

Deregulation in Sweden

In Sweden a new electric power law came into force on 1 January 1996. The power market was deregulated, or reregulated, as some prefer to say, since there still is a legal framework that regulates power production and the market. With this new law the monopolies for sale of power were abolished and power trade was opened to competition. Every customer can now freely choose from which power company to buy electric power.

Production, trade and distribution of power are strictly separated. These tasks used to be managed by the same company – municipal utilities, state-owned Vattenfall or private power companies – that had sole rights to sell and distribute power within their regions of operation.

According to the new law, trading in power is handled by power trading companies, and the distribution of power by grid operators. Many power companies and utilities have created separate affiliated companies to handle power trading and grid operation separately. These are companies within parent companies; with the requirement that their bookkeeping and accounts are separate, so that no money can be moved from grid operation to power trade and vice versa.

Grid operation is still a monopoly, since it is considered too expensive and irrational to install competing power grids. The Energy Market Inspectorate issues licenses to grid operators. These give sole rights to distribute power in specific regions. Prices for this service have to be reasonable, and are supervised by the Energy Market Inspectorate. The price for power is however completely liberalized and is only regulated by supply and demand in the power market.

An electric power customer always has to make two separate contracts, one with a power trading company for the purchase of power and another with the grid operator which distributes the power to the end user. The customer pays one charge to the grid operator for the distribution of the power, and another to the power trading company for electric power. In addition there is an electric power tax and green certificate fee. Finally 25 per cent value added tax (VAT) is added to all these charges.

The grid fee constitutes about 20 per cent of electric power bills, and the charges for power and for taxes around 40 per cent each. The only cost that can be changed by choosing another power trading company is the price for the power itself, which constitutes less than half of the power bill.

The Swedish certificate system

In Sweden a system of green certificates (in Sweden called *elcertifikat* – electric certificates) was introduced in 2003 to promote development of renewable energy. Renewable energy sources, like wind power, biomass, hydro power, etc., will compete with each other on equal terms, and the price is set by supply and demand in the certificate market. The cost of the certificates will be paid by consumers.

Energy-intensive industries are excluded from the obligation to buy certificates; however, they can earn and sell certificates if they have some renewable power production, even if the industry uses this power within their own factories.

Power plants that produce renewable electric power have to be approved by the Energy Agency. These power plants can register to get certificates:

- all wind turbines (irrespective of size)
- solar power from PV cells
- power plants using biomass as fuel
- small-scale hydropower plants (< 1,500 kW)
- new hydro-power plants
- retrofits of existing hydro-power plants
- CHP plants using peat as fuel.

When these producers of renewable electric power have been approved and registered, they will get one certificate for each MWh/year (1,000 kWh/year) from the state. Certificates will be issued for 15 years. The Swedish National Grid (Svenska Kraftnät) monitors the production of the power plants.

All power-trading companies are obliged by law to have a specific share, *quota*, of renewable energy in the power they are selling to customers. They can obtain this share by buying certificates from power producers. The cost for these certificates is then passed on to the end users. The size of the quotas is decided by the parliament and to increase demand for renewable energy the quotas are raised from year to year. If political targets are not met, the quotas can be modified to increase demand and thus certificate prices.

There is a penalty for power trading companies that do not fulfil their quota obligations. Certificates are issued annually, but don't have to be sold in any specific year. Norway was integrated in the green certificate system from the beginning of 2012.

Nord Pool – the Nordic electric power market

The Nordic countries, Sweden, Denmark, Norway and Finland, have a common electrical power system, and also a common marketplace – Nord Pool, which is managed and operated by Nasdaq OMX.

There are six different actors on the market:

- *TSOs.* Owners/operators of the high voltage transmission network, with overall responsibility for the electric power system
- *Producers.* Owners/operators of power plants
- *Traders.* Companies that buy power from producers and sell it to consumers
- *Brokers.* Companies that negotiate business deals between producers and traders
- *Consumers.* End users of electric power
- *Grid operators.* Measure and report power consumption.

How they interact is illustrated in Figure 5.6.

The price on the Nord Pool spot market is set by the marginal cost of the most expensive power that has to be used to satisfy demand. Power producers submit bids for how much power they can deliver and at what price 24 hours ahead to Nord Pool.

The lowest bids, usually from hydro power and wind power with low marginal costs, are taken first, and then the more expensive bids, until the demand is covered. All producers get the same price, corresponding to the highest price for power that will be used (see Figure 5.7).

The price of power depends on several factors. On the supply side, the short-term climate has a great impact; for hydro power it is the annual

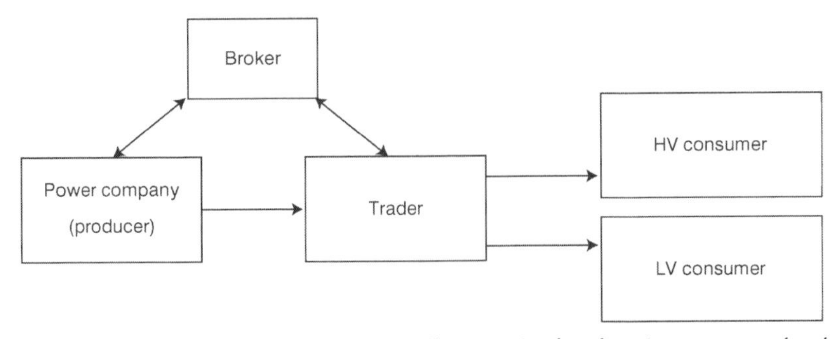

Figure 5.6 Model of the trading system. All power in the electric power market has to pass through a trading company. The power producer sells the power to a trader, who sells it to the end user. Sometimes the prices are negotiated through a broker. The contracts can be based on the hourly spot price on Nord Pool, on monthly average price on Nord Pool, or 1–3-year contracts often based on Nord Pool futures (expected price)

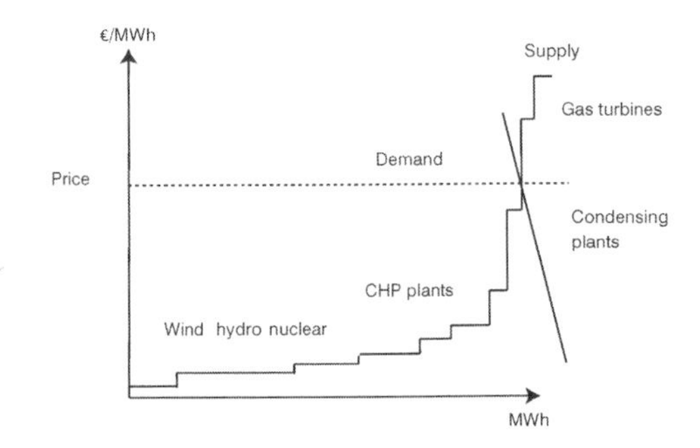

Figure 5.7 Supply and demand curve for electric power exchange. When demand increases, the price of power will increase too. This follows from the merit order system, where the power plants with lowest costs are used first (wind, hydro and nuclear) and plants with higher costs (CHP plants, condensing plants, gas turbines) are added until supply equals demand. All producers will then get this price (dotted line) for the power produced and delivered (Pöyru 2010)

precipitation, which varies significantly in different years. For wind power it is the real-time wind that matters, and much effort is being made to develop good wind prediction models. Fuel prices – and in the European Union also prices in the Emission Trading System (ETS) – have an impact on the power price from thermal power plants.

On the demand side it is also the weather – the temperature and the wind – that determines how high demand will be, especially when electric power is used for heating. The international economic situation and demand for power from industry plays an important role as well. Power producers also have the option of selling power in other markets if they have access to them. Power consumers do not have a similar option (see Figure 5.8).

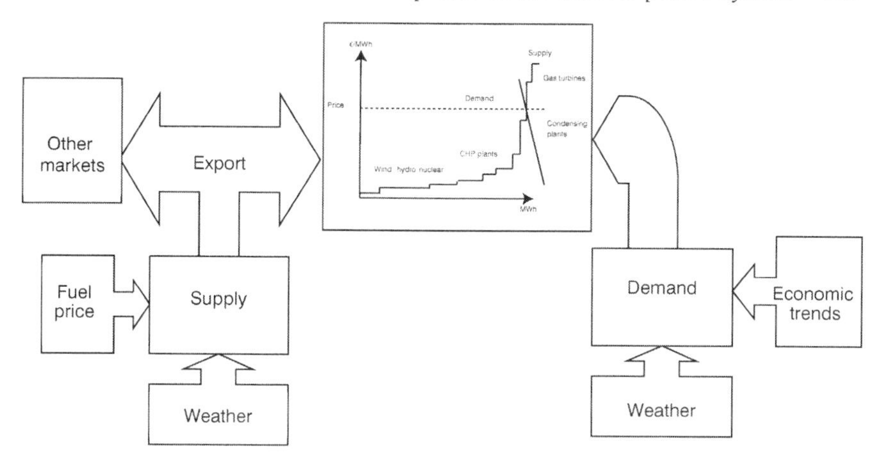

Figure 5.8 Factors with impact on Nord Pool electric power price. The demand for electric power varies according to regular diurnal, weekly and seasonal patterns. On top of this, global economic trends and variations in weather conditions have impacts on demand. On the supply side, weather conditions have a direct impact on the production from wind turbines and PV cells, and seasonal variations on hydro power. World market fuel prices as well as price levels on export markets for electric power from the Nordic countries have an impact on the supply to the Nord Pool market

Regulating power market

If the scheduled power production (supply) does not fit the actual power demand (predicted 24 hours in advance), power production has to be regulated (increased or decreased). To handle this need there is a *regulating power market*.

Balance responsible players (BRPs), i.e. power companies or large consumers who have a balance agreement with a TSO, bid the regulating power to the regulating power market, and the TSO determines the use of reserves (and regulating power market) according to net imbalances of the (Nordic) system. If there is a lack of power in the system, power companies that have made bids to the regulating power market can supply this, and the power company that has the lowest bid (price) will be ordered to do that. If there is a surplus of power, the power company gets paid for reducing the power output. After the operating hours, the net imbalances of all BRPs are calculated and charged for according to country-specific rules.

Electric power market

The rules of the game at the Nord Pool market are described in detail by Ivar Wangensteen from Trondheim University in *Power System Economics: The Nordic Electricity Market* (Wangensteen, 2007).

He describes the power market before and after deregulation. The book's focus is on economics and describes the different models used to calculate prices on the Nord Pool spot market, balance market, etc. The certificate system is also described and analysed. The focus is on the Nordic power market; most cases and examples are taken from Norway, but it also covers power system economics in general.

Electric power as a physical entity and as an economic commodity were separated by deregulation. Electric power was to be treated as any other commodity, and prices set in a market place reflecting the balance between supply and demand.

However, Wangensteen points out that electricity has some features that differ from most other commodities:

- *Real time balance.* Electricity is produced and consumed simultaneously and continuously; at all times there must be a balance between supply and consumption.
- *Not storable.* Electricity cannot be stored in significant quantities in an economic manner.
- *Consumption variability.* The consumption varies, with patterns for day/night, weekly and annually.
- *Production variability.* The production of some renewable power plants varies as well. Wind power varies with the wind speed, and the production potential of hydro power with annual and seasonal precipitation.
- *Non-traceability*. There is no way to track a unit of electric power back to a specific power plant.
- *Customer dependency.* Electric power is essential to households, companies and most other activities. It cannot be disconnected without grave consequences and huge economic losses.
- *Overall responsibility.* In all power systems there must be a TSO which has the overall responsibility for the electric power system.
- *No real-time market.* There can be no real-time market for electricity; prices are either set ex-ante or ex-post.

Wangensteen uses the classical market price model, with price on the *y* axis and quantity on the *x* axis. There is a supply curve and a demand curve, and the price is set where these curves meet, at the so-called Marshallian supply–demand cross.

This is called a state of simultaneous equilibrium. It represents, according to this theory, an optimal solution for society, and maximizes economic efficiency and social welfare. However, this happens only when *perfect market conditions* exist.

Preconditions for perfect competition

Perfect market conditions mean that there are many independent and competing producers, and that consumers have access to all information about the market. To obtain a perfectly competitive market, these conditions have to be met:

- Each market participant must be too small to be able to affect the market price, i.e. all market participants must be price takers.
- All market participants must be economically rational; producers maximize profits, consumers maximize their utility.
- All market participants must have perfect knowledge about prices etc.
- There has to be free entry to the market.
- There are no transaction costs.

In a perfect market the price should, according to theory, reflect the *short-term marginal cost* (STMC). This gives maximum economic surplus, if there are no external costs. There are however some sources of inefficiency. The so called *x*-inefficiency is internal inefficiency in production due to outdated technology, overstaffing and bad management. Market inefficiency means that the price does not reflect marginal cost and marginal willingness to pay.

Another source of inefficiency is imperfect competition due to *market power*. If a participant in the market is large enough to affect the market price, it can increase its profits. The simplest version of imperfect competition is *monopoly*. With a monopoly, producers can control the price by offering different quantities to the market. The producer is then not a price taker, but a price maker. The producer will regard the price as a function of the quantity produced.

If there is more than one supplier in the market, there is still a possibility that some of them can affect the market price. It is described by the *oligopoly model*, developed by the French economist Antoine Cournot in 1838.

The model is based on two preconditions:

- The producers know the demand curve.
- All producers' supply is governed by a given (known) supply from other producers.

When these preconditions are fulfilled, it is easy for all producers to optimize their profits.

According to Wangensteen the Nordic power market is an *oligopoly*. There are a few large producers, Vattenfall, Fortum, Eon, Statkraft, and Dong, and these companies also have a cross ownership, i.e. they own significant shares in each other. They have a common interest to keep electric power prices high.

Consumer response

On every free market, demand declines when prices rise. However, the response from consumers (demand) may be fast or slow. The relation between demand and price is called *price elasticity*.

Price elasticity, i.e. the consumers' sensitivity to price changes, is low on the electric power market. This may vary among different consumer categories and countries. There is also a difference between short-term and long-term elasticity. Short-term price elasticity is *not possible* since power bills are paid as monthly average prices; to reduce power consumption when hourly prices are high does not pay.

When it comes to long-term elasticity it is difficult to get reliable data, since there are so many factors involved. There are however some obvious trends when it comes to the change from oil to other heat sources in the 1980s, due to rising oil prices, and during the last few years the change of electric radiators to heat pumps, which reduced the demand for electric power. These are examples of long-term price elasticity, i.e. a reaction to increased prices for electric power and expected higher prices in the future.

Market prices

Since deregulation, power prices on the electric power market have been rather unstable. For the first years after deregulation, prices went down and stayed on a low level up to the year 2000. Since then prices have been increasing, but in 2012 there was a sharp decrease (see Figure 5.9).

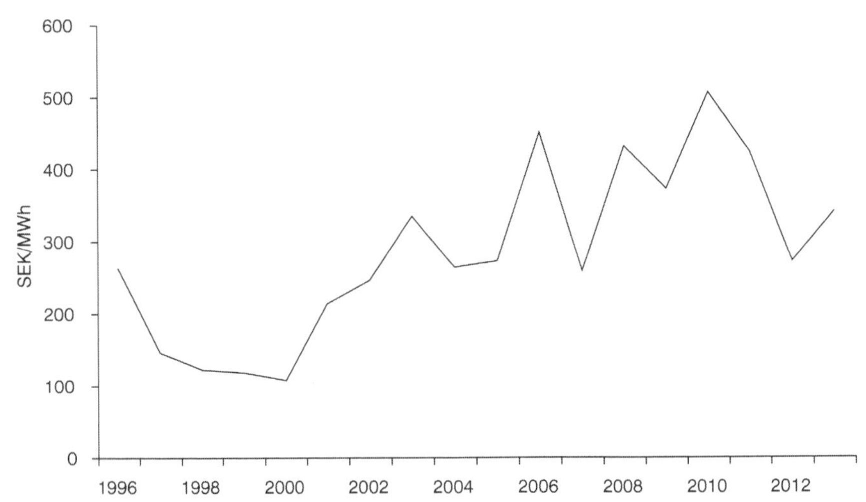

Figure 5.9 Annual average prices on the Nord Pool spot market 1996–2013, SEK/MWh. Average annual prices have varied from 100–500 SEK/MWh, but volatility within each year has been even bigger

Figure 5.10 Prices for green certificates 2003–2013, SEK/MWh. Average annual prices have varied from 200–300 SEK/MWh, but volatility within each year has been much bigger

Certificate prices have also been volatile. The justification for introducing the certificate system was to replace direct subsidies financed by the state budget (by taxes) with a market-based system that would be paid for by consumers. However, the value of the certificates is set and governed by politicians, since the size of the quotas (demand) and maximum price (penalty) are decided by parliament.

The certificate system can, if it works, be used either to stimulate or to limit the development of renewable energy and wind power. The system has, since it was introduced, been revised several times, when the quotas for coming years were changed to increase demand and prices. Still prices have been very volatile (see Figure 5.10).

Wind power producers get their revenues from two markets with highly volatile prices. When the power prices and the certificate prices are low simultaneously, it is difficult to make ends meet. Investments in wind power have a high risk.

The impact of wind power on market prices

Although wind power needs some economic support to be competitive, the impact on power prices from wind power is the opposite; more wind power in the power system will lower the spot prices in the power market. This is caused by the merit-order effect (MOE). This effect has been studied in many scientific papers, which have been reviewed by the consultancy company Pöyru, for the European Wind Energy Association.

The conclusions are that an increased penetration of wind power will reduce wholesale spot prices, and also that *consumers will pay lower prices for electric power*. According to another study wind power also reduces intra-day variability of wholesale electricity prices on the Nord Pool spot market.

Remaining barriers

The laws that regulate electric power management vary in different countries. Each country has its own laws. In the EU there are however some directives and recommendations that should apply to all member states. Before they come into force, they have, however, to be transformed into national laws. The EU has taken a decision to deregulate the electric power system, and to enable free power trading across national borders. Some countries have already deregulated their power markets, while other countries hardly have started this process.

In a deregulated market, power production, power trading and power distribution should be completely separated; that is, performed by different companies independent of each other. In many countries however, huge and vertically integrated power companies (that produce, distribute and sell power), often owned by the state and with a monopoly in the market, still remain.

In the deregulated market all independent power producers have the legal right to have their power plants connected to the power grid, if that is technically possible. In several countries, the power companies that operate the grid are still very reluctant to let independent power producers connect wind turbines to their grid. This is sometimes prevented by the introduction of very strict technical requirements on wind turbines. The European Wind Energy Association has described the situation in this way:

> Grid codes and other technical requirements should reflect the true technical needs and be developed in cooperation between independent and unbiased TSOs, the wind energy sector and independent regulators. ... Grid codes often contain very costly, challenging and continuously changing requirements and are developed in highly non-transparent manner by vertically integrated power companies, who are in direct competition with wind farm operators.
>
> (EWEA, 2005:90)

The capacity of the grid can be a limiting factor for development of wind turbines. For large projects it will sometimes be necessary to reinforce the grid, which takes a quite large investment. The question whether these investments should be made by the project developer or by the grid operator is a matter of constant controversy, at least in countries where there are no strict rules on this in the law.

References

Blomqvist, H. (2003) *Elkraftsystem*. Stockholm: Liber.

EWEA (2005) *Large scale Integration of Wind Energy in the European Power Supply*. Brussels: EWEA.

Holttinen, H. (2004) *The Impact of Large scale Wind Power Production on the Nordic Electricity System*. Helsinki: VTT Publications.

Pöyru (2010) *Wind Energy and Electricity Prices: Exploring the 'Merit Order' Effect*. Brussels: EWEA.

Söder, L. (1997) *Vindkraftens effektvärde*. Elforsk Repport 97:27. Stockholm: Elforsk.

Wangensteen, I. (2007) *Power System Economics: The Nordic Electricity Market*. Trondheim: Tapir Academic Press.

6 Project development

Developing a wind power project includes many different steps that can vary depending on the preconditions: planning, acquisition of consents, agreements and contracts, financing, installation and finally operation of the wind power plant.

During the feasibility study, the developer will have to decide after each step if it is worth continuing, or if it is better to end the project at an early stage and find a better site to develop. The requirements of authorities have also to be fulfilled so that necessary permissions will be granted, and documentation of the estimated production strong enough to convince banks and investors.

The prospects for development of a wind power plant depend on the outcome of economic calculations. If the preconditions are good enough, the wind power plant should be designed to optimize efficiency and output, and at the same time minimize impacts on the environment (see Figure 6.1).

Project management

How a wind power project should be managed depends on who is in charge of the project. If a large corporation plans to invest in wind power, the management of the corporation will give the task of developong the wind power plant to a technical consultancy firm – an experienced wind power developer. Another option is to order a turnkey wind power plant from a developer or a manufacturer, and then give the same company the task of operating the plant. The responsibility of single contractor is very clear: he is responsible for delivering a turnkey wind power plant including the wind turbines, foundations, access roads and grid connection.

If a company's business strategy is to develop and operate wind power plants, it will manage the project with its own staff, and engage external experts and subcontractors as necessary. It will use project financing and has to negotiate loans from banks and/or raise equity. If a new company is formed for the wind power development, the partners have to select a suitable business model, a project manager and CEO, and raise seed capital to hire experts and finance the venture until the first project has been developed and sold.

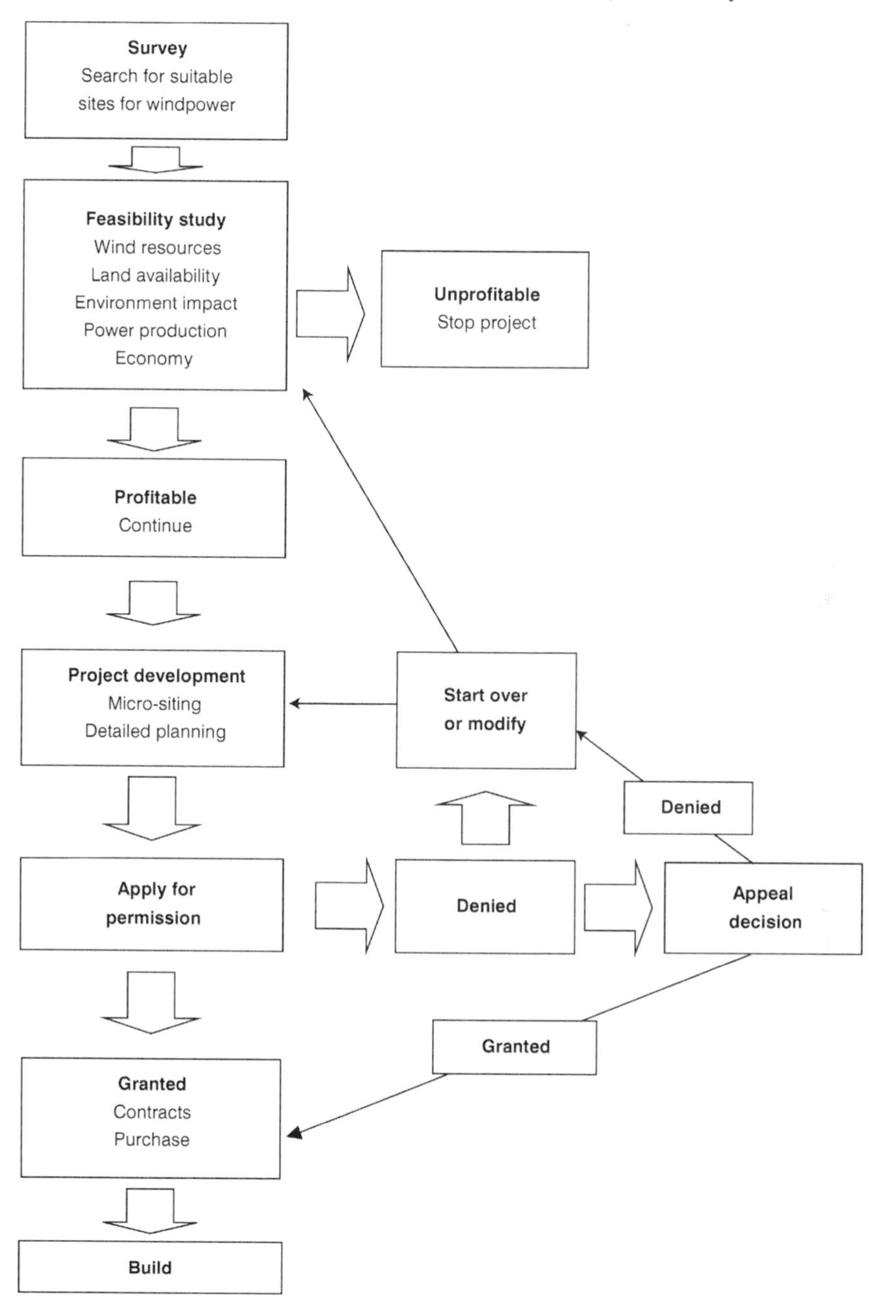

Figure 6.1 Project development process

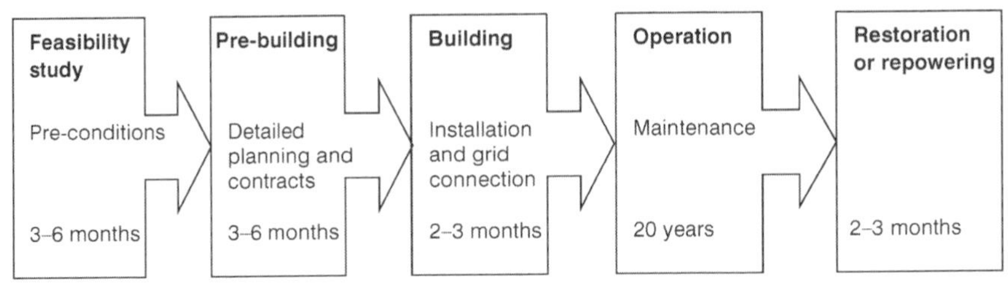

Figure 6.2 Development stages of a wind power project

Managing the project development is a complex task. A detailed project plan and timeline has to be drawn up. First, a feasibility study has to made, to find out if the project will be viable. When the decision to go ahead has been taken, the development consists of three phases: pre-building, building and operation (see Figure 6.2).

Finding good sites for wind power

Whether the task is to develop one, or a few, wind turbines or a large wind power plant within a specified geographical area – a country, region or municipality – the first step is to make a survey of the area to find suitable sites, and then choose the most promising sites for feasibility studies.

The most important precondition for a good wind power project is that there are good wind conditions at the site. The first step is always to study wind resource maps for the area, if any are available. If there are no maps, information about wind conditions can be discovered by analysing data from meteorological stations.

It is long-term wind conditions, the regional wind climate, that has to be found and evaluated. This means the average wind speed for at least a 10-year period, the frequency distribution of these wind speeds, and also if possible the quality of the wind – the turbulence intensity.

The most important aspect is, of course, the wind resource. However, also local conditions such as hills, buildings and vegetation influence the wind and have to be considered in a more detailed calculation of how much energy wind turbines will be able to produce at a specific site.

The wind turbines have to be transported to the site and connected to the grid. The distance from existing roads and/or harbours, the cost of building access roads, ground conditions that influence the design and cost of the foundation, and the distance from and capacity of the grid also have to be considered in the evaluation of a site. When the wind turbines have been installed, they should not disturb people living nearby. Rules and regulations about the maximum acceptable noise level (in dBA) define the minimum distance from buildings in the vicinity of the site.

Permission from authorities is also necessary to install wind turbines. Each country has its own rules and regulations for granting permissions.

The authorities usually check that wind turbines will not interfere or create conflicts with other interests or enterprises. It is therefore necessary for a wind power developer to check what opposing interests there may be at a potential site. It can be air traffic (turbines are quite high), military establishments (radar, radio links, etc.), nature reserves, archaeological sites, and so on..

A good site for wind power development is defined not only by the available wind resource, but also by available roads, power grid and the absence of strong opposing interests.

Feasibility study

When a site with apparently good wind resources has been identified, the first thing is to verify and specify the wind resources.

Wind resource maps are made with a coarse resolution, often a 1 × 1 km grid, so the wind data are smoothed out and can't be used to calculate the production of wind turbines at specific sites. But there are other methods for doing this, such as the wind atlas method. For larger projects it is often necessary to measure the wind at the hub height of the planned wind turbines, using a wind-measurement mast; it is, however, not installed until the feasibility study has come to the conclusion that it is worthwhile trying to realize the project. Wind data are also necessary for economic calculations, and the institutions that will finance the project often require this measurement to be done on-site. A wind atlas calculation can be used to evaluate a site as a first step in the feasibility study.

Then other preconditions for wind power have to be examined:

- *Land.* Who owns the land in the area? Are there landowners willing to lease land for wind turbines?
- *Grid connection.* Is there a power grid with capacity to connect the wind turbine(s) within a reasonable distance?
- *Opposing interests.* Are there any military establishments, airports, nature conservation areas or other factors that could stop the project?
- *Permission.* Is the chance of obtaining necessary permissions reasonably good?
- *Neighbours.* Noise and flickering shadows should not disturb neighbours. Can the turbine(s) be sited so that these disturbances can be avoided?
- *Local support.* Does the local community support development of wind power in this area?

Land for wind power plants

In an agricultural area local farmers usually own the land. In that case, it is quite probable that it will be possible to find landowners who are prepared to lease or sell some land for installation of wind turbines. The land can be farmed as before, but there will be additional revenues. Not only the land, but also the wind can be harvested; to make money out of air is usually considered a sound business idea. In other cases companies, local authorities or the state can own the land and information on land ownership can be found in the land registry. Often landowners make contact with developers to have wind turbines on their land.

Access to land is necessary to be able to install and operate wind power plants, so an agreement with the landowner(s) should be made at an early stage. If several landowners are involved, a common agreement should be made, although the land lease contracts will be individual. Land lease contracts can already be signed during the feasibility study with a paragraph included with the precondition that the agreement will only come into force when the project is realized.

Grid connection

Power lines are usually indicated on maps, so it is quite easy to estimate the distance from the turbine(s) to the grid. However, it is also necessary to know the voltage level, since that sets a limit on the amount of wind power (MW) that can be connected to the power grid. There are several technical factors to take into consideration (the dimensions of the lines, voltage level, power flows, distance to the closest substation, loads, etc.), and only electric power engineers can make these kinds of calculations.

There are however some rules of thumb that give an idea of how many MW of wind power that can be connected to power lines with different voltage levels. One such rule is that grid connection capacity increases with the square of the voltage level (when voltage level is doubled, wind power capacity can be increased four times). Around 3.5 MW can be connected to a 10 kV line, and 15 MW to a 20 kV line, 60 MW to a 40 kV line, etc. Close to the transformer station, more wind power can be connected than close to the end of a power line.

There are also technical rules, so-called grid codes, but there are no harmonized rules on an international level. To get this information correct, it is best to consult the grid operator.

Opposing interests

There is a possibility that the realization of a project will be stopped by so-called opposing interests. The first thing to check is if there are any military establishments close to the site that could be affected by wind turbines. Military establishments with radar or signal surveillance, radio communication links and similar equipment are secret, so they won't be found on maps. The developer should make contact with the appropriate military personnel to enquire if they will oppose wind turbines at the site. If they will, the project will not proceed. In such cases, the developer could ask the military to suggest a site that will not interfere with their interests.

Wind turbines are high structures and can pose a risk to air traffic, especially if there is an airport close by. There are strict rules on how high structures close to the flight paths to and from an airport may be. These rules are available from national aviation authorities. There are also rules and regulations for warning lights for air traffic, which depend on the height of the turbines.

In many countries there are areas which are classified of national or international interest for protecting nature or cultural heritage, such as national parks, nature reserves, bird protection areas, etc. In these areas, and sometimes also in the vicinity of these areas, it will be difficult to get the necessary permissions for wind turbine installations. Protected areas are usually indicated on public maps.

Permission

Spending time and money on projects that can't be built is a bad business and evaluating the prospects of getting the necessary permissions from the relevant authorities is thus a very important part of the feasibility study. The developer has to be familiar with all the rules and regulations that apply to a wind power project and how the authorities interpret them. If there are any local or regional plans with designated areas for wind power development, these will give a good idea of the chances of getting the necessary permissions granted.

Impact on neighbours

To eliminate the problem of neighbours being affected, the turbines should be sited a minimum distance of 400 metres from the closest dwellings. For a large wind power plant this distance may have to be increased. The site where the turbines will be installed should be quite large and have an open terrain. A good rule of thumb is to have a minimum distance of 400 metres for single turbines, or four times the total height (hub height plus half the rotor diameter) if the turbines are very large, and a few hundred metres extra for wind power plants with many turbines. At these distances the impacts from noise should be well within acceptable limits. During micro-siting, more exact calculations should be made of impacts of noise and also shadow flicker on neighbours.

Local support

The attitude of the local community to a proposed wind power project in their vicinity is largely dependent on how the developer performs. In Europe, according to opinion polls and experience, most people have very positive opinions about wind power. On the local level however, there always seem to be some people who strongly oppose wind turbines in their neighbourhood.

How the local community reacts often depends on how they learn about the project. If they get proper information at an early stage, most of them will be positive. When the developer has decided to realize the project, it is important to create a dialogue with local authorities as well as the public, and to take the opinions of the local community about distance to dwellings and other practical details into serious consideration. When the turbines are on-line it is valuable to have local support as people will keep an eye on the turbines and report when problems occur.

There are, however, also dedicated opponents to wind power. Even if these opponents are few, they can cause delay, increase costs and even stop projects that are planned by appealing against the building and environmental permissions given by the authorities. This makes it even more important to give proper and good information to all who will be affected by wind power projects. Making efforts to give information in the local language, and creating some local benefits for those who will live close to the wind power plants, like work opportunities, dividends to local councils or other local organizations, is money well invested. This will prevent the residents in the vicinity feeling concerned and exploited by the project developers.

Project development

When the site where the wind power plant will be installed has been identified in the feasibility study, the exact number and location of the turbine(s) has to be decided. Usually there are several factors to consider: how much power can be connected to the grid; specification of minimum annual production; maximum investment costs; and requirements of economic return of investors/owners. The developer's task is to plan an optimized wind power plant within the limits of given conditions and restrictions. The first task during the pre-building phase is to confirm and specify the details of the feasibility study. All the assumptions made should be re-examined and justified to avoid expenditure on non-viable projects.

In many countries the only permission needed, up to a certain size of project, is a building permit from local government. For large projects, there can be a requirement for permission or licence from higher levels – regional or even national government. There is also a risk that permission will not be granted. This sets a limit to how much should be invested during the pre-building phase.

Land lease

How large an area that will be needed depends on the size and layout of the wind farm. Limits are set by the capacity of the grid and the size of the

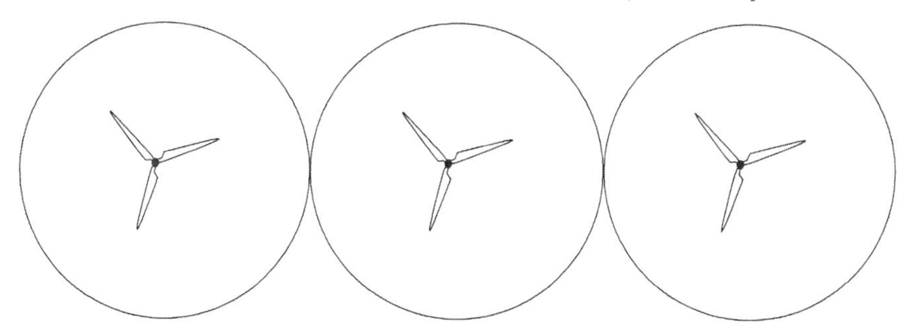

Figure 6.3 Distances between wind turbines. A distance circle with a radius of 2.5D (rotor diameters) can be used to ensure that the proper distance between the wind power plants is 5 rotor diameters

project. Estimating how many circles with a radius of say 2.5 rotor diameters (for an in-row distance between turbines of 5 diameters) can fit into an area without overlap will give the number of wind turbines that can be installed on a given piece of land (see Figure 6.3).

The land used for a wind power plant is called a wind catchment area. For a group of three wind turbines with 64 m rotor diameter and with in-row distances of five rotor diameters, the wind catchment area would then be around 13 hectares.

The project developer has to sign land lease contracts with all the landowners within the area needed for the wind farm. The terms of a land lease contract is a matter of negotiation between the landowners and the developer. In Sweden, the lease usually is set at 3–4 per cent of the gross annual income from the wind power plant. In the United States, the annual land lease usually is in the range of $2,000–4,000 per MW installed.

It is wise to make a fair deal that is in accordance with other similar contracts. Another point is to sign land lease contracts with all landowners within the *wind catchment area*, or the *project area*, if the project is developed on several estates with different owners (see Box 6.1).

Landowners can of course also develop and operate their own wind power plants. In countries such as Denmark, Germany, Sweden, the United States and Canada many farmers own and operate wind turbines sited on their own land.

Micro-siting and optimization

The developer's task is to optimize the wind power plant within the limits set by the local conditions. To find the best solution, wind turbines of different size (hub height and rotor diameter) and nominal power should be tested (theoretically) at several sites within the area. Production should be calculated for these different options and the economics analysed. The impact on neighbours and environment should also be checked. Finally the developer has to choose the best option.

Box 6.1 Wind catchment area

Wind turbines just need a few square metres of land for the foundations, access roads and sometimes transformers or substations. When a wind power plant is operating, the land can be used as before the wind turbines were installed for agriculture or pasture.

The wind turbines, however, need more space in the air: a sphere with the same diameter as the rotor. If several wind turbines are installed, the distance between them should be around five times the rotor diameter. A wind turbine needs a wind catchment area which can be defined by a circle around the turbine with a diameter of five rotor diameters (see Figure 6.3). Land lease contracts should be made not only with owners of the land where the foundations are built, but with all landowners within the wind catchment area.

Another way to define the project area, i.e. the area for which land lease contracts should be agreed on, is to draw the border along the sound emission limit, which in Sweden is 40 dBA. This border is calculated by sound emission software and drawn on a map. The reason for using this border is that neighbouring landowners can no longer use this land to build new houses, as the max allowed sound emission at dwellings is just 40 dBA (in Sweden). This sets the limit for the intrusion of wind turbines.

These principles and definitions are not regulated by law, but it is good practice to use models and contracts which avoid conflicts with neighbours.

In practice there are always *boundary conditions* to consider. Dwellings (minimum distances to avoid disturbance from noise and shadow flicker), buildings, woods, roads, power grid, topography, land property borders, coastlines, etc. define these conditions and limit the area available for siting wind turbines.

With the aid of know-how, good judgment, a constructive dialogue with neighbours and authorities, and high-quality wind data and wind power software, the developer will establish the best solution for the project: a detailed plan that can be realized.

Verification of wind resources

Reliable data on the wind resources at the site are essential and are necessary to make the project *bankable*. They are also necessary for the optimization of the wind power plant. A full year of wind data at various heights, including hub height, is required to be able to establish the wind profile. Installing a couple of 80–100 metre high meteorological masts for very large projects is quite expensive. If there is any doubt about the outcome of the permission process, project developers postpone these measurements until permissions are granted. This, however, delays the project, so the timing of these measurements is a matter of corporate strategy in evaluating risks and benefits.

There are, however, other cheaper and quite reliable methods of verifying wind resources. If the terrain is not too complex and there are weather stations within reasonable distance – and even better a number of wind turbines that have been operating in the same region for a number of years – the wind resources at a site can be calculated and evaluated by using wind atlas software. There are also new types of wind measuring equipment installed at ground level: sodar which uses sound impulses and lidar which uses light (laser beams), to measure the wind.

Environment impact assessment

In many countries it is compulsory to make an environment impact assessment (EIA) for large wind power plants. The environment impact assessment is a process: a public dialogue. It results in an EIA report, which is evaluated by the authorities in deciding if the project will be given permission to build the wind turbines. Often it is necessary to engage external consultants to conduct the EIA itself, and to make special reports on birdlife and other impacts.

Public dialogue

The developer can start by making rough outlines for the wind power project, and invite people living 1–2 km from the site for a preliminary dialogue at an information meeting. The developer can give information about wind power in general, the environmental benefits, local wind resources, possible impact and finally present an outline of the project and ask the audience for their opinions.

The developer should also have a preliminary dialogue with local and regional government, county administration, the grid operator and other relevant authorities in separate meetings. The project should at this stage be presented as a rough outline, the point of an early dialogue being to keep an open door, so that the project can be adapted and modified to avoid unnecessary conflicts.

In many countries wind power developers have used a planning processs which is in accordance with the intentions of the EIA process. Most developers organize local information meetings at an early stage to ensure that the public are well informed, and hopefully support the plans. Some developers also offer people living in the area the opportunity to buy shares in the wind power plants or offer other local benefits (see Box 6.2). The information meeting, as dialogue with the public, is also the first step in the EIA process.

Box 6.2 Local benefits

A major local benefit from a wind power plant is, of course, local ownership, so that revenue from the power production remains in the region where it has been produced. Since wind power development started some decades ago, local ownership has been quite common in Denmark, Germany and Sweden. Now, as new large power companies have begun to develop wind power projects, this may change. In Denmark legislation was introduced in 2008 to support local ownership; local and regional investors are guaranteed to be allocated ownership of 20 per cent of onshore wind power plants at the cost price of the project.

In Sweden there is no such law, but it is a common recommendation from the farmers' organizations and the Swedish wind power association that, for wind power plants with four or more wind turbines, 20 per cent should be offered to local investors (as in Denmark).

With wind power plants built on land with local ownership, 0.5 per cent of the gross revenues should be paid into a fund for local benefits, to be used for local development. If the land is not locally owned, 1 per cent of gross revenues should be paid to the local benefit fund.

In Sweden it has been common practice since the early 1990s for locally owned wind power companies and cooperatives to pay a dividend to the local community.

The developer has to present several different options for the siting of wind turbines, and also discuss practical matters of the construction process, building of access roads, power lines, etc. Also the so-called zero option – the consequences of the project not being built – has to be outlined. The developer can, of course, argue for the preferred option, but should be sensitive to the opinions that are put forward. The fact that the local community know the area they live in very well has often proved to be useful for the developer.

Through this dialogue, the project is made concrete and is designed to minimize impacts on the environment and neighbours. After this dialogue is completed, the EIA document will be compiled and it will form an appendix to the permission application.

Appeals and mitigation

When the relevant authorities and political or official bodies have processed the application, the developer will eventually get the necessary permissions granted. Appeals against the decision have to be lodged within a couple of weeks. When that time has passed the decision becomes final and the actual building of the windpower plant can start.

It may, however, happen that some neighbours, interest groups or even a public body will raise an appeal against the decision. The developer then has to wait until the court has made a decision on the appeal. These legal processes can delay a project for years and sometimes also put a stop to it. This is another good reason to inform all concerned parties, adapt the project to avoid nuisances, even if it reduces the economic results a little. If the permissions are appealed against, the costs will be much higher.

Micro-siting

When a good area for a wind power plant has been identified, land lease contracts are signed, and the prospects of getting the necessary permissions seem good, the project has to be specified in detail. The number and size of wind turbines, and their exact position, have to be defined. A wind power plant can be configured in many different ways, but there is often an optimum way to do it that will maximize the return on the investment. This fine-tuning of a wind power plant's layout is called *micro-siting*.

Energy rose

An energy rose is the best guide in minimizing the impact of wind wakes from other turbines. A regular wind rose shows the average wind speeds and the frequency of the wind from different directions. An energy rose shows the *energy content* of the wind from different directions. Since it is the energy in the wind that is utilized by wind turbines, this is the best guidance. An energy rose can be created with wind atlas software (see Figure 6.4).

Wind power plant layout

Small groups with two to four wind turbines are often put on a straight line, perpendicular to the predominant wind direction. The distance between turbines is measured in rotor diameters, since the size of the wind wake depends on the size of the rotor. A common rule of thumb is to site the turbines five rotor diameters apart if they are set in one row. Larger wind power plants can have several rows of wind turbines. In that case the distance between rows usually is seven rotor diameters (see Figures 6.5 and 6.6).

This ideal model for the layout can be applied in open, flat landscapes and offshore. The actual layout of a wind farm is, however, often formed by the limits set by local conditions, such as land use, distance to dwellings, roads and the power grid. If there are height differences on the site, this will also influence how the turbines should be sited in relation to each other to optimize power production. It is usually not reasonable to increase the distance between turbines to eliminate the impact from wind wakes completely as this is an inefficient use of land.

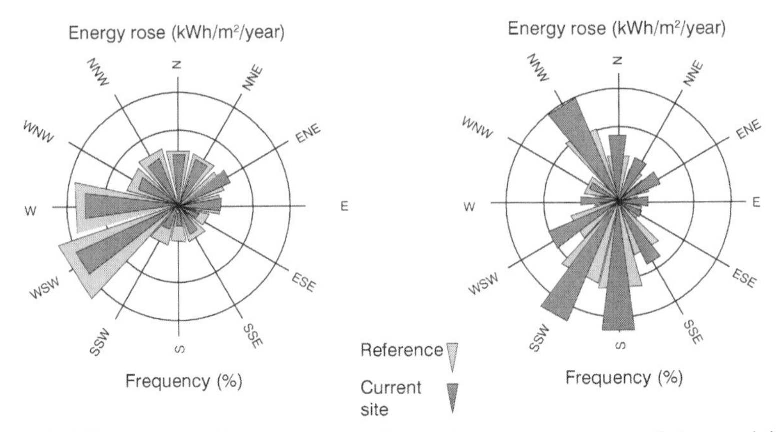

Figure 6.4 Energy rose. An energy rose shows the energy content of the wind from different directions. A) In the left-hand diagram (south coast of Sweden) most energy is in the winds from WSW and W. A line of wind turbines should then be installed in a line running NNW–SSE. The in-row distance can be quite short. B) In the right-hand diagram (an island in the northern Baltic sea) energy comes from more directions, but it shows that the line of turbines should be oriented from W–E. Both are very windy sites, close to open sea

In areas where one or two opposing wind directions are very dominant the in-row distance between the turbines can be reduced to three or four rotor diameters (see Figure 6.7).

In large wind power plants with several rows of wind turbines, the in-row distance should be five rotor diameters and the distance between rows seven rotor diameters. In offshore wind power plants these distances ideally should be six in-row and eight to ten between rows. Wind wakes survive longer at sea, since the turbulence over water is lower. It is the turbulence in the surrounding wind that destroys the wakes (see Figure 6.8).

If the area isn't absolutely flat, the optimal configuration will be irregular with distances between turbines differing and the turbines not set in straight lines. In practice the layout is also guided by aesthetic and practical concerns: along a coastline, road, headland, a regular pattern or in an arc as in the offshore Middelgrunden wind power plant outside Copenhagen (see Figure 6.9).

Optimization

A project developer should, of course, optimize the layout of a wind power plant. It is, however, important to be aware of which parameters should be optimized. For the owner and operator of the wind power plant, it is neither the installed power in MW nor the total power output that should be optimized (maximized), but the *cost efficiency*. It is the production cost for each kWh of electric energy that should be minimized. At the same time the available project development area has to be utilized in an efficient way. For

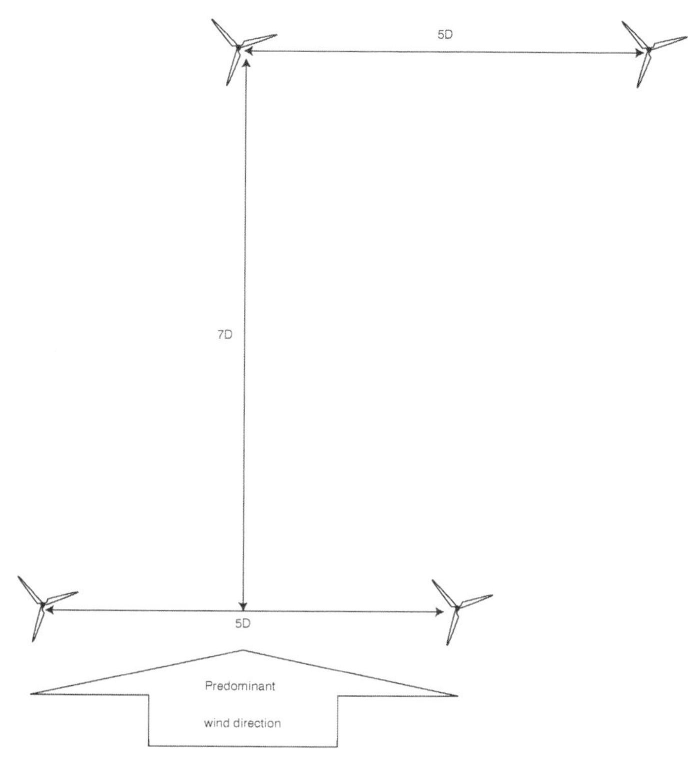

Figure 6.5 Wind power plant layout. The rule of thumb is to have an in-row distance of 5D (rotor diameters) and a between-row distance of 7D

Figure 6.6 Wind power plant in Falkenberg, Sweden. This wind power plant has rows which are perpendicular to the predominant wind direction. The in-row distance is 5D and the distance between rows 7D (photo: Falkenberg municipality)

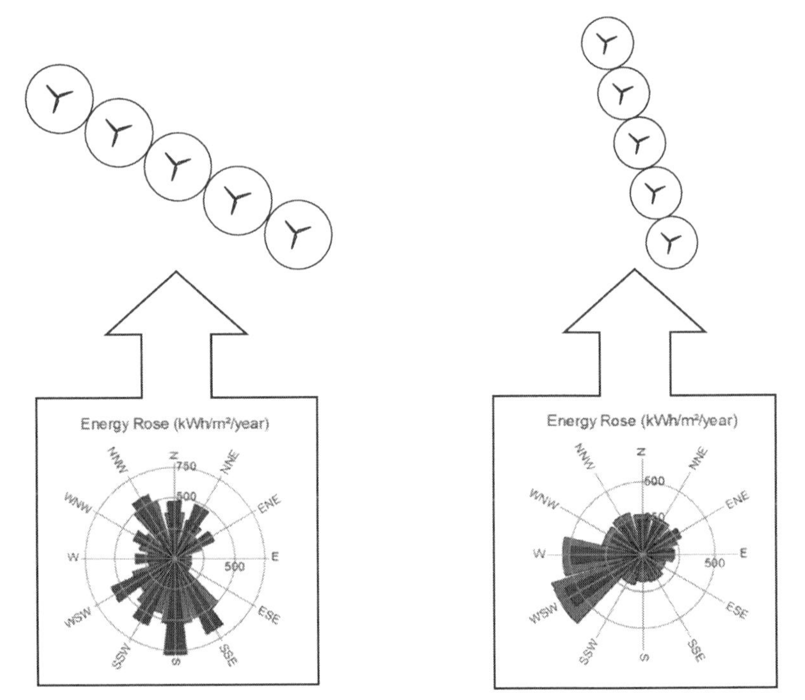

Figure 6.7 In-row distance. The standard in-row distance is 5D (rotor diameters), as in the illustration to the left. If the wind mainly comes from one and the same direction (or from two opposite directions), as in trade wind regions, the distance can be reduced to 4D, illustration to the right. If several rows are installed the in-row distance should be increased to 4–5D, and the distance between rows to 8–10D

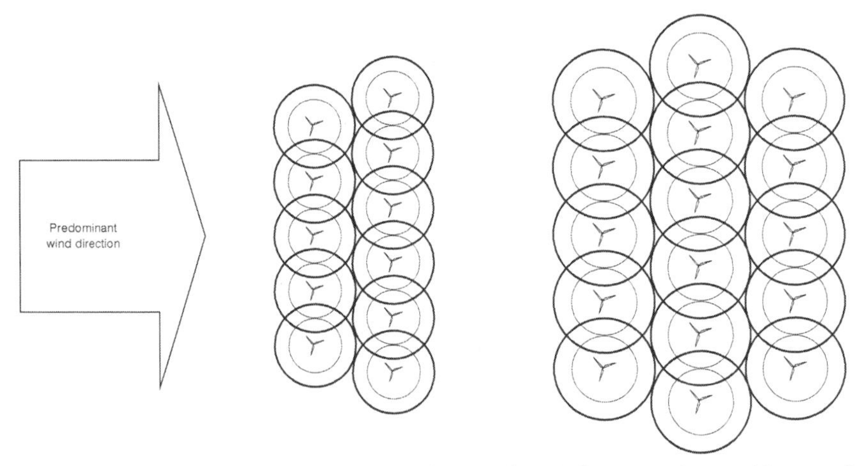

Figure 6.8 Layout of large wind power plant. Left: wind power plants with several rows of turbines, the in-row distance should be five rotor diameters (small circles), and the distance between rows seven rotor diameters (large circles). Right: offshore the distances should be six diameters in-row and eight to ten between rows.

Figure 6.9 Middelgrunden offshore wind power plant. The wind turbines in the offshore wind power plant Middelgrunden just outside Copenhagen are sited no more than three rotor diameters from each other, which is much too close for optimal production. In this case aesthetic concerns weighed heavily as these wind turbines are visible from the Danish parliament building (photo: Tore Wizelius)

the landowner and for the wind turbine supplier as well, it is most profitable to install as many turbines as possible on the land. The owner/operator has the choice of optimizing the rate of return on investment, or the cash flow generated by the wind power plant.

The available land area and the capacity of the grid restrict the maximum power that can be installed. The developer, or rather the customer who has ordered a wind power plant, may have restrictions when it comes to the total investment cost. It is the relation between these factors that will set the framework for the optimization of the wind power plant configuration. The wind turbines themselves should also be tailored to fit the wind resources and other conditions at the site.

Park efficiency

Park efficiency is the key concept in the optimization process. When many turbines are installed at the same site, the wind turbines will 'steal' some wind from each other. How large these *array losses* will be depends on the configuration of the wind power plant, i.e. the positions and distances (in rotor diameters) between the wind turbines.

Park efficiency is defined as the relation between the actual production of a wind power plant to what the production would be without any array losses caused by other turbines. The closer the turbines are sited in relation to each other, the lower the park efficiency. To aim at a park efficiency of 100 per cent is not realistic, it would be bad use of available land. It should, however, be as high as possible. With the same number of turbines in the same area the park efficiency can be optimized by following the rules of thumb described in Figure 6.7, but also by fine-tuning the position of turbines and checking the park efficiency by calculations using wind power software.

Looking at costs, a park efficiency of 90 per cent (i.e. a loss of 10 per cent of production) can be compensated by distributing the necessary investment for access roads, grid connection, cranes etc., to more wind turbines. However, the wind wakes that reduce output will also increase wear and tear on the wind turbines, since turbines will be exposed to more turbulence from the wind wakes of other turbines. Installing wind turbines too close to each other is bad practice, and optimizing the technical lifetime of the wind turbines should also be included in the cost-efficiency calculation.

Access roads and internal grid

When the final layout of the wind turbines is decided, it is time to plan access roads, work areas, internal grid and grid connection. The dimension and other characteristics of access roads are specified by turbine manufacturers. Wind turbines are very large and heavy and have to be transported by special vehicles. Next to the site, areas have to be cleared and prepared for unloading of rotor blades, nacelle and tower, as well as for a crane to lift these components into position. Often the rotor is mounted on the ground before it is lifted up to the nacelle by crane. The construction of access roads can be simple or very complicated, depending on the character of the terrain.

There are different options for configuration of the internal grid. This task is undertaken by a consultant – an electric power engineer. Cables are usually laid in the ground in ditches along access roads, or else integrated into the road in cables reinforced by pipes strong enough to withstand the heavy loads of trucks.

Conflicting projects

When several project developers operate in the same area, there may be conflicts of interest. If a new wind power plant is installed close to existing wind turbines, especially if it is in the prevalent wind direction, it will reduce the power output of the existing wind power plant. This is, however, a matter for the planning authorities. A distance of 15 rotor diameters or more is recommended in such cases. Often there are planning regulations that set a minimum distance, usually a number of kilometres, between wind power plants. Another option is to cooperate and form a joint venture and to develop a larger wind power plant.

The capacity of the grid is another delicate question. There is always a limit to how much power can be fed into the grid. If too much capacity is connected, the wind turbines have to be cut out of the grid when they all produce at full power. This is a situation that should be avoided. It is up to the grid operator to regulate how much power can be connected, and if the capacity already connected is close to this limit, it is often better to look for another site. Another option is to reinforce the grid, but the question then is if the grid operator or the project developer should pay for this. Rules and regulations about this vary in different countries.

With several projects in the planning stage, and a limited grid capacity, it is important that the grid operator has transparent rules for which projects will have the right to connect wind power plants to the grid.

Estimation of power production

To estimate how much a wind turbine will produce at a given site, two things have to be known:

- the power curve of the wind turbines
- the frequency distribution of the wind speed at hub height at the site.

A power curve shows how much power a turbine will produce at different wind speeds. It is shown as a table, graph or as a bar chart and is available from the manufacturers. These power curves have been verified by independent authorized agencies. Power curves are, however, valid only under specified conditions in an open landscape. If turbulence is too high or the wind gradient exponent too large, production will be reduced. This is a risk especially in forest areas or close to forest edges.

It is necessary to have detailed information about the winds at the site. It is not sufficient to know the annual average wind speed. It is also necessary to know the frequency distribution of the wind speeds, i.e. how many hours a year the wind speed will be 1, 2, 3, ... 30 m/s. These data should represent the wind speed distribution during a normal year, i.e. average values for a 5–10-year period. The data also have to be recalculated to the hub height of the turbines. Then the power produced at each wind speed is multiplied with the number of hours this wind speed occurs. There will however always be some losses as the wind turbines have to be stopped for regular servicing; some power is needed to operate the turbines, and there are losses in cables, transformers, etc.

Wind data sources

The data on wind conditions at a prospective site can be obtained from several different sources, and preferably from a combination of them. The sources for wind data can be of four different kinds:

- historical meteorological data
- satellite data
- on-site measurement data
- data from meteorological modeling.

Historical meteorological data

National meteorological institutes have collected data on winds for many decades and have a lot of wind data in their archives. Wind measurements have been made on many different locations, so there is a wide geographical spread of data. There are very long time series of wind data. However, these data are rarely from representative sites, since the main interest in the wind conditions have been at sea, harbours and airports. The standard height for measurements, 10 metres agl, is also quite low. Still, these archived data are very valuable for reference, and for the calculation of long-term wind conditions. The basic data are usually public and free of charge; for special data treatment and time series there can be charges.

Satellite data

Satellite data are available from NCEP/NCAR, where wind data sets from 1948 up to now in a 2.5 degree longitude/latitude grid are available, or from MERRA (Modern-Era Retrospective Analysis for Research and Applications), with a finer resolution. Satellite data are used mainly for long-time correlation of measured wind data.

On-site measurement data

A wind-measurement mast at the prospective site will collect the most accurate data on wind conditions. Ideally the measurement mast should have the same height as the hub height of the wind turbine(s). Since the cost increases with height, this may be prohibitive for smaller projects. The measured data can in such a case be recalculated to hub height. The best data will be collected at the top of the mast, since the mast itself affects the wind. The wind should however be measured at two or more different heights, making it possible to calculate the wind gradient exponent.

There are many good wind-measurement masts with equipment to collect wind data for wind power plants available, which are quite easy to install at the site. Cup anemometers measure the wind speed, and wind vanes the wind direction. Also temperature and air pressure should be recorded. The data are sampled and recorded by a data logger, which has to be very robust and well protected from rain. The data can be collected remotely. But data losses can occur due to power failure and water ingress. Measurement data should cover one year, to get wind data for all seasons.

New types of equipment have also come into use: sonic detection and ranging (sodar) and light detection and ranging (lidar). These are installed on the ground, and send sound pulses (sodar) or light pulses (lidar) up into the air. Sodar and lidar can get measurement data not only from a point, but from a three-dimensional space. Sodars/lidars have not replaced measurement masts, but are often used as a complement, to get data from additional heights, nearby sites in an area or to get data on the turbulence in complex terrain. Sodars/lidars are easy to transport and install, and less expensive to use than high measurement masts. Some, but not all, financial institutions accept data from sodars and lidars as good enough to make projects bankable. This is logical, since data from modern sodars/lidars are very reliable, and probably even better than data from wind-measurement masts.

All the relevant factors can be calculated from these measurement data – the average wind speed, the frequency distribution of the wind – and these can also be specified for different wind directions. They can be transformed to different heights and also the turbulence intensity can easily be calculated from these data.

Since all the data cover only a limited period of time, often one year, which might not be representative for an average year, the data have to be correlated to long-term wind data as a basis for calculating the expected power production of a wind power plant at the site. Such long-term corrected data should be used for all calculations, including turbulence, etc.

Data from meteorological modeling

On most sites it is possible to calculate the power density and energy content of the wind without using measuring equipment at the specific site. Instead the wind data from measuring masts at other sites, within 10–50 km distance from the site, can be used. These data come from measuring masts used by the meteorological agencies which also have historical long-term data. These wind data can be recalculated to represent the wind climate at the site where wind turbines will be installed. These calculations are made with the wind atlas method.

There are also so-called meso-scale models available, but these are very complicated and are handled by specialists. These models are mainly used to create wind resource maps for countries or regions, but can also be used with higher resolution for specific sites.

Long-term correlation

The winds can differ much from year to year and the measurement period could have been exceptionally windy or calm. To find out if the data collected during one year are representative for an average year, these one-year data have to be compared to long-term wind data. To do this it is necessary to have a reference site, at a site with the same wind climate, where the wind has been measured for several years.

The measured wind data have to be compared with corresponding data from the same measurement period in the same region, where long-term data are also available. Then it can be checked how representative the data from the measurement period are, compared to the long-term data from the second measuring mast. National meteorological institutes have collected wind data for decades from a large number of meteorological stations in different parts of their countries. Nowadays satellite data are often used for long-term correlation of wind data. Finally the collected wind data can be adjusted so that they will correspond to a normal year – the long-term average.

If the wind speed is measured with a wind-measurement mast at a site for a shorter period, for example for 6–12 months, the wind energy during a normal year can be calculated by using wind data from a nearby measuring mast with long-term data available if there is a correlation between the wind at the two sites.

Usually these data are available from an official meteorological station in the same region. If not, there may be useful satellite data available.

There are several ways to correlate measured data to predict the wind for a period of years. These so-called measure-correlate-predict (MCP) methods can be used for long-term correlation:

- regression
- matrix
- matching Weibull parameters
- wind index.

With the regression method the wind data are entered into a graph to check if they correlate well enough. Then a transformation function based on the concurrent data is used to convert the reference data to the conditions on the site of the local measurement, and finally the long-term data from the reference site are transformed to site data.

Planning tools

When it comes to actual planning, there are many very good tools available which make the process easier. These programs are basically geographical information systems (GIS) with many special features developed for wind power project development. These tools perform all the calculations needed.

There are several different computer programs for wind power applications that are based on the wind atlas method (see Box 6.3).

In complex terrain and where available data are unreliable (in mountain areas, large lakes and at sea), this method can't be applied and it is necessary to make on-site wind measurements. For large projects, banks and other

Box 6.3 Wind Atlas Software

WAsP has been developed by Risoe National Laboratory in Denmark and is the basis for all wind atlas programs. It can be used to make wind resource maps, wind atlases for whole countries, as well as production calculations for single wind turbines or large wind power plants.

The program WindPRO can do the same calculations as WAsP and has additional modules for noise, shadow and visual impact, planning tools and many other functions, as well as a comprehensive database with wind turbine models and wind atlas data for regions and countries. It has been developed by Energi og Miljödata in Denmark.

The program WindFarm that has been developed in the UK by the company ReSoft, and WindFarmer from GL Garrad Hassan, can do all the calculations necessary for project development, including optimization and visualizations. There is also a freeware called RETScreen which can be found on a website developed by CANMET Energy Technology Centre in Canada, with education, databases and software for different renewable energy sources.

All of these programs are easy to work with and give reliable results if the operator understands them and is an experienced user. They can be used to calculate how much a wind turbine of a specific brand/model can produce at a given site, as well as the sound propagation, park efficiency and visual impact. It is also possible to create wind resource maps with some of these programs.

financial institutions will also require wind data from a meteorological mast installed at the site.

In many countries the state-owned meteorological institute has prepared wind atlas data for some hundreds of measurement masts in different parts of the country. Wind atlases with wind atlas data are available for most countries in Europe and for many countries outside Europe; many of them are available on the internet.

Roughness of terrain

The wind is retarded by friction with the ground surface. How much depends on the character of the terrain where the wind turbine will be sited and on the surrounding landscape. To calculate how much energy a wind turbine at a specific site can be expected to produce, wind data from one or several measuring masts within a reasonable distance from the site are used. These data (giving mean wind speed and frequency distribution for an average year) have to be adapted to the specific conditions at the chosen site – the roughness of the terrain. The roughness is classified into five different classes (see Table 2.1).

How much a wind turbine can produce depends not only on the character of the terrain at the site, but is influenced by the terrain over a large area. The terrain conditions close to the site have the greatest impact on the turbine's production. The roughness usually varies in different sectors and thus with the wind direction. In calculations an area with 20 km radius around the turbine site is divided into 12 sectors, 30 degrees each, with the wind turbine in the centre. A roughness classification is then done, sector by sector, or by setting roughness values to areas of different character directly on a map.

The classification of the area close to the site should always be made in the field, since maps do not give an exact picture of reality; symbols for buildings, roads, etc. are larger/wider than they actually are so as to make them visible on the map. Significant changes could also have occurred since the map was made (new buildings, roads, etc.). For distances more than 1,000 metres the classification can be made at the desk from a map or directly in the software (see Figure 6.10).

Hills and obstacles

If a wind turbine is sited at the top of a hill or on a slope this could increase its power production. The speed-up effect from hills has most impact at lower heights above the hilltop. The height of this effect increases with the size of the hill (see Figure 2.8).

Steep slopes, however, can have the opposite effect; if the inclination is larger than approximately 25 degrees, the slope can create turbulence which will decrease production. If the surface is rough or complex, this could happen with inclinations of 10–20 per cent and if the slope is covered by trees or forest, there will be no hilltop effect if the slope is more than 5 degrees. A wind atlas program will calculate the impact of hills and obstacles on production. In a complex terrain it is always necessary to make wind measurements on site, not only to get correct wind data, but also to measure the turbulence.

Buildings and other obstacles close to a wind turbine (less than 1,000 m away) affect the wind that the turbine will use. How much an obstacle affects the production depends on the height, width and distance from the turbine and its character (porosity). Buildings and other obstacles that are situated 1,000 m or more from the site should not be classified as obstacles, but as elements in the roughness classification.

If there are large obstacles close to the turbine, production will be affected. For large wind turbines the impact from obstacles are comparatively small, since the impact depends on the difference between the turbine's hub height and the height of the obstacle. The turbulence from an obstacle will spread to twice the obstacle's height (see Figure 2.6). The rotor of a

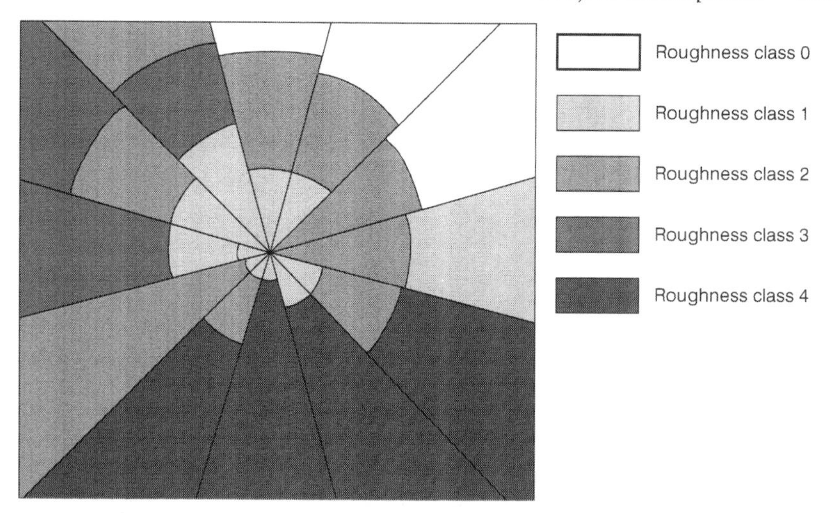

Figure 6.10 Roughness rose. The roughness of the terrain is classified for each sector, and then the production of the wind turbine is calculated for each of these sectors (source: WindPRO2)

turbine with 80 m hub height and 80 metre rotor diameter has its lowest point 40 m above ground level, which means that an obstacle has to be more than 20 m high to cause turbulence within the rotor swept area.

Information about obstacles (within 1,000 m from the site) and height contour lines are entered into the program. The wind speed will change each time the roughness of the terrain changes. The wind atlas program recalculates wind atlas data to wind data at hub height for each sector. This is illustrated by an energy rose, which is used as a basis for the layout of the wind farm.

Fingerprint of the wind

The energy rose is the fingerprint of the wind, at a specific height. When a wind atlas calculation is made, it will produce tables and diagrams where the energy produced by the wind turbines at the site is specified for wind directions and wind speeds. These calculations are made before the wind turbines are installed; and by analysing the energy rose and other data from these calculations, the configuration of the wind power plant can be optimized by moving the turbines around until the best results are achieved. In different countries and regions these fingerprints differ, as is shown by these two examples from the west coast of Sweden and Sri Lanka respectively (see Figures 6.11 and 6.12).

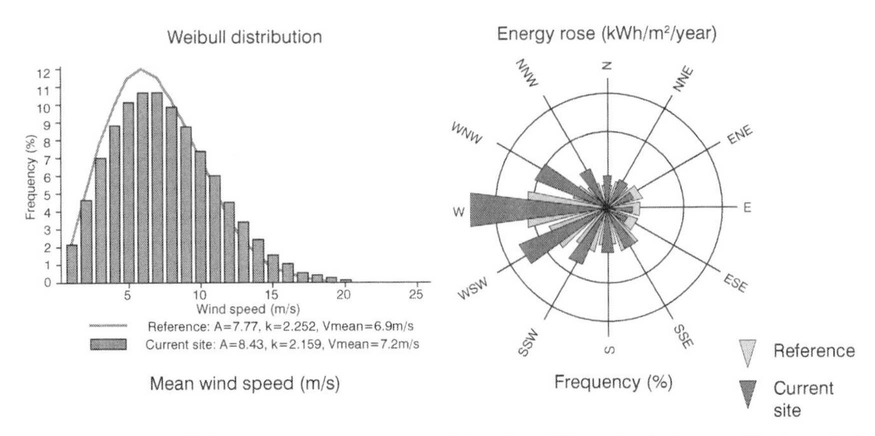

Figure 6.11 Wind fingerprint: west coast of Sweden. The calculations with the wind atlas software WindPro for the site that has been classified in Figure 6.10 shows how much energy is produced by the wind turbine from different wind directions. The reference shows the wind energy with roughness 1 at the site

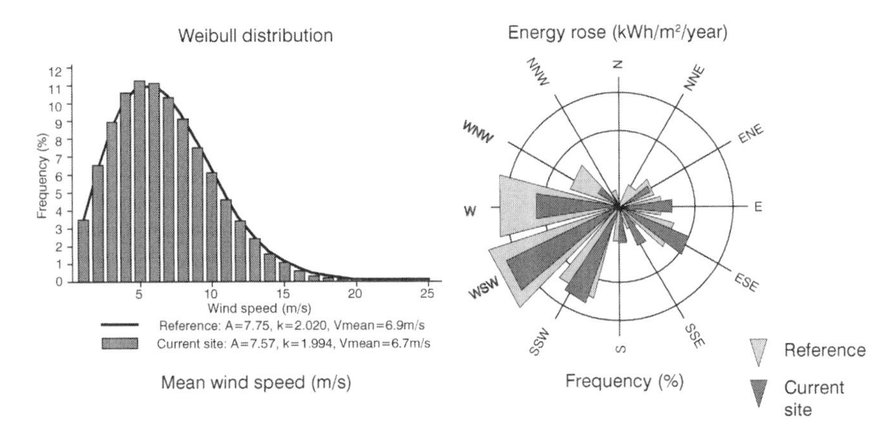

Figure 6.12 Wind fingerprint: west coast of Sri Lanka. In Sri Lanka, with a monsoon climate and trade winds, the energy rose shows that almost all energy is produced by winds in the sector SSW–W

Wind atlas calculation

First the input for the project, maps and height contours for the area are loaded. Then roughness is defined and wind turbines entered on the map. Often meteorological data are available in the program in a database, as well as wind turbine models. This wind atlas software then calculates the estimated production of the wind turbines, park efficiency and there are even tools that can find the most efficient configuration that will optimize the production of a specified number of turbines within a limited area. There are also modules that calculate noise levels in noise sensitive areas, and shadow flicker.

The wind atlas program first calculates the wind's frequency distribution at hub height for each sector, and then multiplies the frequency distributions with the wind turbine's power curve. The results are weighted according to the frequency for each wind direction and finally summarized.

If the terrain is not extremely complex this method gives very accurate results. However, to make accurate calculations with this method, it takes quite a lot of practice and experience of how different kinds of terrain in a region should be classified, and experience of how far from the measurement mast wind atlas data are still representative.

If there are other wind turbines on-line in the area, production of these can be entered as a reference in the calculation. If these turbines have been in operation for a couple of years, and their average production is known, these figures can be entered in the program. Then a calculation is made with the software, and the results can be compared. If the actual production is the same as the calculated, then the calculation for the turbines in the project should also be correct. If not, the quotient of actual/calculated production can be multiplied by the calculated production of the new turbine to make it more accurate.

If a wind-measurement mast, sodar or lidar at the site has collected wind data, these data can also be entered in the software. Such on-site data will give the best estimation of the power production. With these kinds of data, and after adjusting them to long-term wind data, a local wind resource map can be created which shows how the wind will vary within the site, depending on the character of the local terrain, heights and obstacles.

Sources of error

The accuracy of the calculation depends, of course, on the quality of the data that are entered into the program. Wind atlas data are based on measured data from different periods. Which sequence of years the data are based on is indicated in the database and can vary for different measuring masts. These periods can be too short or not be representative for the long-term averages. The wind data as well as the transformation of these to wind atlas data can be impaired by faults due to technical faults of the measuring equipment, etc. or systematic errors when the data were recorded. There is a certain amount of rounding when data are transformed to Weibull parameters that are used in the software.

The roughness classification is never absolutely correct, and the roughness can change with the seasons and during the lifetime of the turbine. The power curve of the turbine is a third source of error. The form of the power curve depends on the conditions when it was measured, and does not give an exact relation between wind speed and power. In a different surrounding with different terrain and wind regime it may vary somewhat from the certified power curve. Special care should be taken in areas with tree cover and forests.

These and other factors are considered to create an error margin of 10 per cent in the calculations, an estimation confirmed by experience. From a statistical point of view, the uncertainty with the wind resource estimation can be either too low or two high. However, most of these factors are can only be negative, caused by losses in the internal grid, the turbines own power consumption, etc. Therefore 10 per cent should always be subtracted from the result of the calculation.

Instead of using the default reduction of 10 per cent, losses and uncertainties can be calculated in more detail with wind atlas software (see Box 6.4).

Wind measurements

Even if the wind atlas calculations give accurate results when the assumptions for the calculations are valid, it is often necessary to also get wind data from a meteorological mast installed at the site, with one full year of data. This is often a requirement from financing institutions or prospective buyers of the

Box 6.4 Loss and uncertainty

Seven different loss categories, with some sub-categories, have been defined and agreed on. The starting point is the gross annual energy production (AEP) without any losses, and array losses are also excluded. From this starting point losses are deducted to get an estimate of the actual power production.

Wake effects

The wake loss category has two subcategories: wake effects for all wind turbines (park efficiency, array losses); and future wake effects (impacts from wind turbines planned but not yet built).

Availability

Availability losses have four subcategories: turbine (losses due to maintenance, etc); balance of plant (faults in internal grid up to the substation); grid; and other.

Turbine performance

Turbine performance losses have four subcategories: power curve; high wind hysteresis (losses due to shutdowns between cut out and subsequent restart); wind flow (turbulence, high wind shear, etc.); and other.

Electrical

The electrical loss category has two subcategories: losses (up to the interface to external grid); and facility consumption (reductions of sold energy due to consumption for operation of plant, behind the meter).

Environmental

The environmental loss category has six subcategories defining impacts on power production from climate and nature: performance degradation not due to icing (blade soiling and degradation); performance degradation due to icing (reduced aerodynamic efficiency of blades); shutdowns (due to icing, lightning, hail, etc,); extreme temperatures (outside the turbine's operating range); site access (due to remote location, weather or force majeure events); forestry (the growth or cutting down of trees in nearby forests that will reduce or increase production).

Curtailment

The curtailment category has seven subcategories: wind sector management (required shutdowns to reduce loads); grid curtailment and ramp rate (due to limitations in external grid); PPA curtailment (when power purchaser does not take power from the plant). There are also curtailments to reduce impacts on environment and neighbours, for noise, shadow flicker, birds and bats.

Other

Finally there is a category, other, for losses not accounted for in the first six categories.

wind power plant. For some sites, far from weather stations or in a complex terrain, there is no other way to get reliable data.

To get correct and detailed information about the power density and energy content of the wind at a specific site, a wind-measurement mast can be installed. Anemometers mounted at different heights record wind speed and wind direction by wind vanes. A data logger collects the wind data. With these data the average wind speed, the frequency distribution, power density, energy content, distribution of wind directions, wind profile and turbulence intensity for the measurement period can be calculated. The ideal is to measure the wind at the hub height of the turbines to be installed, but it is easy to recalculate wind data to other heights. Instead of a meteorological mast, sodar or lidar can be used. These measuring instruments are far cheaper and give accurate results with high resolution.

The precondition for accurate results is, however, always that the measuring equipment works according to specifications. The anemometers have to be calibrated, and installed correctly on the mast. During actual measurement weather conditions can affect the results, especially snow and ice.

The inaccuracies involved in the measured data can be estimated with statistical methods.

Before wind measurement starts, it is important to make sure that the financiers will accept the wind-measurement methods that will be used and that the measurements are bankable.

Pitfalls

Evaluations of wind turbine sites, analysis of wind data, wind power plant configurations and calculations of estimated production, are usually based on standard assumptions. There are however sites where these assumptions do not apply. There are several factors that can upset the predictions that the investments have been based on. Temperature and climate is one factor, extreme wind speeds another, and the impact of trees and forests should also be considered. Wind resource maps should be interpreted critically and finally it is important to be aware of the risk of upgrading a wind farm before it is built without adapting the wind turbine configuration accordingly.

Extreme temperatures

There are many places in mountains and arctic regions with very good wind resources, and in some of these areas in the United States and in northern Scandinavia wind turbines have been installed. Arctic climate conditions put some special demands on the wind turbines. Ordinary standard turbines would not survive for long in these conditions, but they can be adapted to the strains that the climate will cause; special steel for towers, special oils for low temperatures, heating systems for rotor blades and anemometers, ice detectors and other special equipment can be used to increase availability and power production for wind turbines in cold climates.

Extreme heat seems to be less of a problem, since most turbines have cooling systems installed. In tropical climates the humidity in the air can be very high, and may make it necessary to install extra air conditioning equipment inside the hub and where the control system is installed.

Extreme wind speeds

Most wind turbines are designed to survive extreme wind speeds of at least 60 m/s, but only for a few seconds of time, i.e. extreme gusts. But turbines can be designed for sites where the wind speeds can be higher than this and the cost of the turbines will, of course, also increase. Survivability has been tested in practice as severe storms, hurricanes and typhoons have occurred in areas with many wind turbines installed.

In Denmark, most of the thousands of turbines installed survived the severe storms that occurred in the beginning of 2005, storms that felled millions of trees in southern Scandinavia. In India 111 turbines in a wind power plant in Gujarat were uprooted by a hurricane in 1998. In Japan wind turbines on Miyaki island were felled by a typhoon. In August 2006 the typhoon Sangmei destroyed the wind farm Changnan Xiaguan with 28 turbines in China.

In areas where hurricanes and typhoons are part of the normal climate, wind turbines should be designed to survive these conditions, even if it will make them more expensive. Another option is to avoid such areas and disqualify them in the feasibility study.

Forest

Avoiding trees and forests has long been the first commandment for wind power developers. The reason is, of course, that forest has high roughness and retards the wind. Because of this the wind shear will increase and the wind gradient exponent should not exceed 0.2 over the rotor swept area. Trees and forest edges also create much turbulence, and it is recommended to have a safe distance between wind turbines and forest edges. If the distance isn't sufficient, the power curve is not valid; calculations will overestimate the expected production (see Box 6.5).

In Germany and Sweden, wind turbines have been installed in forested areas. These installations are often made on hills and ridges in the woodlands, which often gives a very high production. On such sites the turbines are operating high above the surrounding forests, often where the hill height plus the hub height is several hundred metres. Otherwise there are serious pitfalls when wind turbines are installed in forests.

To avoid a too high wind shear and turbulence, towers have to be very high; the bottom end of the rotor should be above the so-called roughness sublayer, which in most cases ends 30 metres above the treetops. Today many manufacturers offer very high towers 100–140 metres, to make it possible to utilize the wind resources in forest areas.

Developing wind power in forests has many drawbacks:

- higher towers, higher cost
- same wind speed at hub height – *less* energy in forests (due to turbulence etc.)
- high turbulence and shear – more wear and tear, shorter lifetime, higher operational and maintenance costs.

Box 6.5 Wind power and forest

The following rules of thumb can be applied near forest edges:

Shear (wind gradient exponent) should not exceed 0.20.

The height of trees in the vicinity of a wind turbine, from the turbine's bottom flange, should not exceed the following limits, where R is the distance from the turbine to the trees, H_h is the hub height and D is the rotor diameter:

For $R \leq 5D$ Maximum height of trees: $H_h - 0.67D$

For $5D < R \leq 10D$ Maximum height of trees: $H_h - 0.67D + 0.17D \times ([R/5D]-1)$

If a turbine with a hub height of 60 metres and a rotor diameter of 60 metres is going to be installed at a site with a forest edge where the trees are 20 m high [60 – (0. 67 × 60)= 20 m], the distance to the edge should be at least five times the rotor diameter, 5 × 60 = 300 metres. At a distance of 600 metres the max height of trees will be ~30 metres:

$$60 - 40 + 10.2\ [(600/300) - 1],$$

to avoid affecting the performance (the power curve) of the turbine. If the distance is given, say 500 metres, the option in the second case would be to increase the hub height to 64 metres. The first recommendation is thus to have a distance of more than 10D to forest edges, and if that is not feasible, to increase the the hub height according to these rules (Schou Nielsen and Stiesdal, 2004).

Some manufacturers have developed control methods for such sites. One method is sector management, which means that the wind turbines are stopped when the wind comes from directions with too high turbulence, which of course will reduce the potential gross annual production. Avoiding trees and forest is still the best advice. Wind turbines installed too close to forest edges often produce 20–40 per cent less power than expected. Hills and ridges in woodlands can, however, be very good sites for wind power. This is not always the case, so on-site measurements are compulsory for all sites in forest areas.

Wind resource maps

Wind resource maps are usually made with meteorological models, using high-capacity computers. These meso-scale models are usually very good. However, the input data also have to be correct. In specific types of terrain changes, such as coastlines or mountainous terrain, the reality may be too complex to be well described by a model. They usually cover large areas, such as a country. The roughness of the terrain cannot be described in detail, but is taken from digital ground coverage maps based on satellite data.

These data are not always correct, and have sometimes led developers to plan projects that had to be ended after wind measurements had proved the wind resource map had overstated wind speeds. Many developers have spent resources developing projects which turned out to be worthless. It is important to interpret wind resource maps with a critical mind.

Upgrading of wind turbines

Developing a wind power project often takes a long time, especially the permission process. If permissions are appealed against, a developer may have to wait for the permission to start the building phase for several years. During this time the technical development of wind turbines continues. Wind turbines in the planning stage for a wind power project may not be available on the market when the building phase is about to start. The most cost-efficient turbines have become bigger than the ones the developer has got permission to build and used for the wind power plant configuration.

One example of this process is the offshore wind power plant Lillgrund in Öresund, between Malmö and Copenhagen. The project was started by the company Eurowind AB in 1997. The wind power plant was originally planned for 1.5 MW turbines with 66 m rotor diameter. Within the planning area, sites for 48 turbines were identified and applied for. Since this was the first application for permission to build a large offshore wind farm in Sweden, the requirements for investigations from the authorities were immense. However, all aspects asked for – impact on fish, sea bottom plants and animals, water streaming, shipping, radar and many others – were thoroughly researched, which took several years. Finally permission was granted in October 2003.

During this time the first partner who had planned to finance the project was involved in other projects. The whole project was sold to the Swedish state-owned power company Vattenfall in 2004. The turbines the wind power plant was designed for had become obsolete. Vattenfall chose larger 2.3 MW turbines with proven offshore performance from the Danish offshore wind power plant Nysted, and finally choose to increase the rotor diameter from 90 to 100 metres.

However, Vattenfall kept the original layout for the wind power plant. Since the layout pattern is governed by the size of turbines, and the distance between turbines is measured in rotor diameters, this was a radical change of the layout. In the original plan they were following the rules of thumb, described earlier in this chapter, but with larger turbines this was no longer valid. The in-row distance was reduced to 3.3 and the between rows distance to 4.3 rotor diameters.

This resulted in a park efficiency of no more than 77 per cent. When the wind comes parallel to a row (with 4.3 diameter distance) the second turbine will produce 70 per cent less than the first turbine in the row. With the wind across the wind farm perpendicular to the rows (with 3.3 rotor diameter distance), the turbines in the wake of the first one will produce 80 per cent less (Dahlberg, 2009).

This wind farm would actually produce more power with fewer turbines! When the wind blows across the rows (3.3 rotor diameter distance), production would increase if every second row was cut out and feathered, and the distance between turbines would increase to 6.6 diameters, which could be done by sector management.

Even if it would take the engineers quite some time to rearrange the farm layout, and recalculate the dimensions of the foundations (where design depends on depth), and take a few more months to get permission for the new wind farm configuration, this would be nothing compared to the gains in cost efficiency for the wind power plant. Never use an old layout for new and larger turbines.

Choice of wind turbines

After more than thirty years of research and development, and with practical experience from thousands of wind turbines in operation, the wind power industry can be considered to be mature. There are no inferior wind turbines that will break down in a few years, at least not among commercial grid-connected models from well-established manufacturers. There is also a certification system that ensures the quality of the wind turbines. Making an investment in wind power is no more risky than any other investment in a reliable technology.

Many different types of wind turbines have been tested and commercialized over these decades. The type that has proven most reliable, efficient and has had most success in the market is the horizontal-axis three-bladed upwind wind turbine, and most wind turbines in commercial operation are of this type.

There is a wide choice of wind turbines on the market. Most manufacturers offer several different models and these models are often available in different versions. This wide choice makes it possible to install wind turbines that are tailored to match the conditions at a specific site.

What type of turbine should be used at a site depends on many different factors: the wind resource; the roughness of the terrain; the size of the available area; grid capacity; and the purpose of the project. Wind turbine prices and delivery periods are, of course, important parameters to consider.

Wind turbine size

There are wind turbines of several different sizes, where the size is defined by hub height, rotor diameter and nominal power. The main parameter for the definition of size is the rotor diameter, or the rotor swept area. The bigger the swept area is, the more wind can be captured and transformed to power. Hub height and nominal power are secondary criteria. The choice of hub height is determined by the surrounding terrain and the nominal power by the wind resources at the site. As the rotor diameter increases, higher hub heights will be necessary. For sites in forested areas, very high towers have been developed, up to 160 metres.

Type of wind turbine

Within the group of three-bladed upwind horizontal-axis wind turbines, there are several different options. Development has moved to more sophisticated technology, which has increased the efficiency, but also the complexity of wind turbines. The power output of most wind turbines is now controlled by pitch, by turning the blades of the rotor. The revolution speed of the rotor has become variable to increase efficiency.

Since the middle of the 1990s gearboxes have been a problematic component. Considering the loads that gearboxes have to manage, it is not too surprising that many of them have had to be retrofitted or replaced after a few years of operation. Much work has been done to develop gearboxes as well as the drive train, to make them more flexible and durable. There are also models without a gearbox, using a multi-pole direct-drive ring generator. New concepts with a robust one- or two-stage gearbox between the rotor and a slow-running multi-pole generator have also been developed. However, no one would expect a gearbox of a truck to last indefinitely and the same can be said about wind turbines. Consequently, this future replacement cost after some ten years should be included in the economic calculations. The trend since 2010 is, however, that more manufacturers are opting for direct-drive or hybrid turbines.

There are several options also for towers: there is a choice of tubular steel, concrete or lattice towers, and very high towers are often hybrid towers with, for example, an 80 m concrete tower with a 40 m steel tower on top. The choice of towers will, to a large extent, depend on the climate. In northern Scandinavia a lattice tower is not an option, since heavy icing in the winter can cause it to collapse. In tropical countries, such as India, lattice towers have an advantage and since they need regular maintenance – the bolts have to be checked at regular intervals – this is a better option in countries where the cost of labour is low. Lattice towers need less steel, are easier to transport and the foundations also don't need as much concrete as gravity foundations for tubular steel towers.

When a developer has a large area for the installation of a wind power plant, there are also different options, not only for the configuration of the individual wind turbines, but there is also a choice between a large number of small or medium-sized wind turbines, or fewer bigger ones. The biggest wind turbines usually utilize the wind resources the best, both when it comes to power production per land area and cost efficiency. This is, however, not always the case for smaller areas, where it may be possible to install several small wind turbines but only a few big ones, so that the first option results in more installed power and higher production. It is always worthwhile spending time and effort testing different options to optimize the configuration of a wind power plant.

The choice between a small and simple turbine and a huge high-tech one depends on the size of the project, the nature of the site and the conditions in the country. The infrastructure – roads and power grid – is crucial. At some sites a small, simple and robust work-horse of a turbine may be the best choice, while the big high-tech turbines may be the best choice at other sites.

Wind turbines tailored to wind climate

Most manufacturers offer different options of their models. Often two or three different hub heights can be chosen, and there is also a choice between tubular steel, lattice and precast concrete towers. This choice is mainly a matter of taste, cost and transportation to the site. The choice of hub height should be based on the character of the terrain. In an open landscape, at the coast and offshore, the shortest tower will be sufficient, since the wind speed does not increase very much with height.

In other kinds of terrain, with many trees, buildings or close to forest edges, it makes sense and increases cost efficiency to have higher towers. There is a relation between heights and cost; a higher tower is more expensive, so the difference in wind speed has to increase production enough to pay back a higher investment.

Nominal power vs rotor diameter

There are wind turbines designed for sites with low, normal and high average wind speeds. Low-wind turbines have a large rotor diameter in relation to the rated power and have a low rated wind speed (the wind speed where the wind turbine reaches its rated power). High-wind turbines have a small rotor diameter in relation to the rated power and a high rated wind speed. This classification is based on the relation between the nominal power and the rotor swept area A in kW/m^2.

Enercon has two models E-48 and E-53 with 800 kW nominal power. The E-53 is a low-wind version, and E-48 is better for sites with strong winds. The Vestas model V90 with a 90 m rotor diameter has options of 1.8, 2 or 3 MW generators, for low, normal and high wind sites respectively. The Chinese manufacturer Goldwind has a 2.5 MW wind turbine with 90 or 100 metre rotor diameter (see Table 6.1).

The values of these specific ratings (kW/m^2) are in the range 0.28–0.58 among wind turbines available on the world market. The highest values are for offshore wind power plants. To adapt a wind turbine to the wind

Table 6.1 Nominal power in relation to rotor area

Wind turbine model	*Rated power kW*	*Rated wind speed m/s*	*Rotor diameter m*	*Rotor swept area m^2*	*kW/m_2*
Enercon E-48	800	14	48	1810	0.44
Enercon E-53	800	13	53	2205	0.36
Vestas V90-1.8	1800	12	90	6362	0.28
Vestas V90-2.0	2000	13	90	6362	0.31
Vestas V90-3.0	3000	15	90	6362	0.47
Goldwind 90/2500	2500	14.5	90	6362	0.39
Goldwind 100/2500	2500	13	100	7850	0.32

conditions at a site, the specific rating should be around 0.3 for average wind speed of 6 m/s, 0.4 for 7 m/s, and 0.45–0.6 for 8 m/s.

IEC wind turbine classes

There is a system for wind turbine classes established by the International Electrotechnical Commission IEC (www.iec.ch). These classes are used for the standardization and certification of wind turbines. These classes are numbered with Class I for the highest wind conditions to Class IV for sites with moderate winds (see Table 6.2).

Wind turbines are specified for different wind conditions, for example IEC IIB, which means that the turbine may be used at a site with a mean wind speed of more 8.5 m/s, and where the turbulence intensity is more than 0.16

Table 6.2 IEC classes for wind turbines

	IEC class				
	I	*II*	*III*	*IV*	*S*
v_{ref} (m/s)	50	42.5	37.5	30	Values to be specified by the designer
v_{ave} (m/s)	10	8.5	7.5	6	
A I_{15}	0.18	0.18	0.18	0.18	
a	2	2	2	2	
B I_{15}	0.16	0.16	0.16	0.16	
a	3	3	3	3	

The values apply at hub height, and 'A' designates higher turbulence, 'B' designates lower turbulence, 'I_{15}' is the turbulence intensity at 15 m/s, '*a*' is the slope parameter in the normal turbulence model equation at a wind speed of 15 m/s. The turbulence varies with wind speed, so the wind speed has to be specified. Turbulence is measured in 10-minute averages.

The basic parameter used for defining IEC wind turbine classes is the reference wind speed (v_{ref}). A turbine for a reference wind speed v_{ref} is designed to withstand wind climates for which the extreme 10-minute average wind speed, with a recurrence period of fifty years at the turbine's hub height, is lower or equal to v_{ref}. There can actually be higher extreme wind speeds than this and the other measure is the 3-second average for extreme gusts.

Grid compatibility

The condition and capacity of the grid at the site, and the requirements of the grid operator, may restrict the choice of models. When only a single turbine or a small wind power plant is connected, the grid can absorb some spikes, consumption of reactive power and other power quality problems which could be caused by wind power plants without advanced equipment for grid compatibility. For large wind power plants with a significant nominal power and in areas with high wind power penetration, the requirements of grid operators are stricter and limit the wind turbine models that can be chosen and the electrical equipment installed in transformers and substations on the wind power plant operator's side of the grid interface.

Supplier

When a wind power plant has started to operate, it should continue to do so for some twenty years with as few problems as possible. It is therefore important to choose a supplier and wind turbine models that have a good track record. A manufacturer with a good reputation for quality equipment, and who responds quickly to problems that may arise, will help to keep the wind power plant in operation. The best way to find this out is to get in touch with other experienced project developers and wind power plant operators.

When the market for wind turbines is growing fast, it may not always be possible to get orders for wind turbines accepted, especially for smaller projects and minor developers. Delivery times may also be very long. The question then is if it is worth changing the timeline of a project to wait for delivery from the desired manufacturer or to choose another option. Since the technical lifetime of wind turbines is twenty years, it may be worth waiting for delivery from a manufacturer with a proven record, good after-sales service, experience in operation and maintenance, and with spare parts available for the whole period of operation.

There is a wide choice of wind turbines available on the market, and more will come when new manufacturers start to export their products on the world market. Research is an ongoing process, and there are continuing technical developments towards even more cost-efficient wind turbines.

The project developer should never specify a manufacturer's model in the application procedure. It is sufficient to specify the number and size of turbines, hub height and total height. There are several reasons for this. First, the preferred model may no longer be available when the building process is started. Most manufacturers continually develop their turbines and a specific model may have a larger rotor or different standard hub heights when time comes to order the turbines. If so, the permission is no longer valid. Therefore in the application a margin should be added to the size of the turbines. Second, in a business venture several different manufacturers should be asked for tenders.

The project developer needs to have a sound knowledge and understanding of the interplay between site conditions, wind resources and wind turbine specifications to choose the most suitable models for the wind power plants being developed – wind turbines are tailored to the sites.

Economics of wind power plants

A thorough economic analysis is necessary to make a good decision about a wind power project. The wind power plant has to generate enough income to guarantee that the investors, or the banks that provide finance, will get their money back and a decent return on their investments. No investors will be willing to finance the wind power project without a realistic and convincing budget.

The first task in the economic analysis is to make a realistic calculation of how much electric power the wind power plant will produce at the chosen site. This is done with a wind atlas program such as WindPRO, WAsP, or by on-site wind measurements that have been normalized to long-term averages. The results of this calculation will show how many kWh the wind power plant can be expected to produce in a normal wind year; the average annual power production during the lifetime of the plant.

The next task is to estimate the investment costs. When a wind turbine starts to produce power, the income has to cover the costs of finance as well as the costs of operation and maintenance. The wind power plant should also generate a profit.

There are several different methods to calculate the returns on an investment: the annuity method, the present value method, the internal rate of return (IRR) and the payback time method. A cash-flow analysis illustrates the annual returns on the investment during the lifetime of a wind power plant. All these calculations are, however, based on assumptions on future power production (the winds vary), power prices, interest rates, etc. that cannot be accurately foreseen today. The economic analysis should therefore also include a *sensitivity analysis* that will estimate the risks and opportunities with the investment. Finally, a plan has to be made for financing the project that ensures that there always will be enough money available to make interest payments, repayments of loans and other bills. This is called a *liquidity budget*.

Basically the financing can be done in two different ways, *corporate financing* or *project financing*. A wind power project can be financed by borrowing money from a bank, by private investors or in other ways. It also takes money before any income is generated to develop the project, to build access roads, make deposits when wind power plants are ordered. The developer himself or a building loan from a bank usually covers these costs.

Investment

The cost of purchasing the wind power plant, installation at the site and grid connection are estimated in an investment budget. In a feasibility study, rounded estimations can be used. The purpose of the feasibility study is to evaluate if the project is worth realizing or not. After the decision is taken to fulfil the plan, a new more detailed investment budget is made, based on tenders for wind turbines, access roads, foundations, grid connection work and equipment and other ancillary works. This carefully calculated investment budget is then presented to the bank, in the application for a loan, and to prospective investors.

An investment budget has the following items:

- wind turbine(s)
- foundation(s)
- roads and miscellaneous
- grid connection
- project development.

Wind turbines

The price of different models and sizes of wind turbines can be established from price lists or directly from the manufacturers or their agents. During the final procurement (purchase process), prices and conditions can be negotiated. If the wind power equipment is manufactured in a foreign country (with another currency) the price will also depend on the exchange rate, which sometimes can change quite quickly. The transport of the turbine to the site, mounting, installation and grid connection is done by the manufacturer and is normally included in the purchase price. The costs for the mobile crane and some transport costs have to be covered by the developer. For wind power plants installed on land, the cost for the turbine amounts to about 75 per cent of the total investment.

Foundation

The cost of the foundation varies a little between different manufacturers. The price for a rock foundation and a gravity foundation is about the same, while foundations for a lattice tower tend to be cheaper, since less concrete is needed for them. The manufacturer will give the technical specifications for the foundation (size, weight, etc.). The project developer will then ask a local building company to build the foundation. For offshore installations the foundations are much more expensive.

Access roads

The cost of access roads depends on the size and weight of the turbine, ground conditions and the length of the road that has to be built. At the sites where the individual turbines will be installed, it is necessary to prepare an open workspace approximately 50 metres wide and 70 metres long for trucks, cranes etc. In many cases it will be sufficient to reinforce existing roads so that trucks and a mobile crane can get to the site. It is often easier and cheaper to prepare an access road when the soil is hard and dry. After the turbine has been installed, the road only has to carry an ordinary van for the service crew and material. This cost depends on local conditions. The cost of the mobile crane and special transportation costs (for example by ferry) have to be covered by the project developer.

Grid connection

A transformer and a cable or overhead line to the closest grid power line have to be installed to connect the wind turbine to the electric power grid. The cost depends on the size and model of the wind turbine, the distance to the grid and the grid voltage. Large wind turbines have an integrated transformer, either in the nacelle or in the bottom of the tower. If this is the case, the cost of the transformer is included in the price of the turbine. For large wind power plants an internal grid has to be built and often a substation. A telephone line to monitor and control the turbine has to be connected as well.

Project development

This item includes costs of planning – the time the developer has to spend working on the project. This time is needed to prepare an application and EIA, business negotiations, information, economic calculations, etc. Fees for technical consultants, interest payments during construction and other similar costs are also included. This cost can vary a great deal depending on the time needed for the development process and the fees the developer and consultants charge.

Total investment

It is often hard to calculate the total investment cost accurately before the development actually starts. The budget should give a rough idea with contingency for unexpected costs included.

For offshore wind farms the investment costs are considerably higher than for wind power plants on land. The cost for foundations is much higher, as well as for the undersea cables that connect the wind power plant to the grid on land.

Economic result

It is of course just as important to establish the economic outcome after the wind power plant has begun to deliver power to the grid. Then the plant begins to generate revenues, but also incurs some costs. To make a profit the revenues have to be larger than the costs. Calculating future costs is not so difficult, it is far trickier to calculate the revenues.

There are basically two kinds of costs, capital costs (interest and repayment of loans) and costs for operation and maintenance (O&M). The actual capital costs depend on how the project has been financed. There are basically two models for this, *corporate financing* or *project financing*. Corporate financing means that the wind power plant is owned and operated by a company and treated like an investment within the company. It will be included with other machinery in the company's balance sheet. With project financing the wind power plant is treated as an independent economic entity.

If the wind power plant has been financed by loans from a bank, the conditions are specified in the loan agreement. If private investors finance it, for example in a shareholding company formed to own and operate a wind power plant, the project will be financed by equity capital, but the stakeholders expect a good return on their investment. In these cases, financing is made by a combination of loans and equity.

Depreciation

Commercial wind turbines are designed for a technical lifetime of 20–25 years. The actual technical lifetime is not well established, since few wind turbines have reached that age yet. How much retrofitting is necessary when a turbine starts to get older is also a factor of uncertainty. However, nowadays, costs for retrofits during the lifetime of wind turbines are often included in the project budget.

Maintenance costs will, however, increase with age. Therefore the economic lifetime may be shorter than the technical lifetime. After 15–20 years the maintenance costs may be so high that it makes sense to replace a wind turbine with a new and more efficient model. In economic calculations depreciation time is usually set at 20 years, but owners usually opt to pay back loans on a shorter term of 8–12 years. This means that the capital cost (loan plus interest) always should be paid back within 20 years. If the turbine continues to produce power without problems, profits will be higher, unless the income is drained by costs for repairs and retrofits.

Operation and maintenance

A wind power plant starts to operate when it has been installed and connected to the grid, and should do so for at least 20 years. Compared with other power plants, the costs for operation and maintenance of wind power plants are very low as no staff are needed to operate the plants and no fuel has to supplied. However, wind power plants need regular service and maintenance, they have to be insured against accidents and there are also some administration costs.

A wind turbine needs regular servicing, just like any other machine. The service crew will make regular checks of the condition, usually twice a year (depending on the manufacturer). The oil in the gearbox has to be checked regularly and changed after a couple of years. The servicing costs for the first two years are often included in the turbine's price, but oil and other materials are not included. After that initial period the manufacturer or a service company offers a service contract to conduct regular servicing for an annual fee.

During the time of the warranty, which usually will last for two to five years, fire and third-party insurance is necessary. When the warranty runs out, machine insurance is usually added, and the insurance premium will increase.

Wind turbines are often installed on land that is leased, where a land lease contract gives the wind power owner/operator the right to use the land for 25–35 years. The landowner will get an annual fee which is part of the annual operation costs.

Owning and operating a wind power plant also entails some administrative work: invoices and taxes have to be paid and bookkeeping has to be managed.

Revenues

The basic income for a wind power installation are revenues from selling the electric power. The owner has to make a *power purchase agreement* (PPA) with a power trading company that buys and sells electric power. In many countries the power market has been deregulated during the last few years, while in others there is still a monopoly. The conditions for a PPA as well as the price per kWh can vary much in different countries.

There are also some special bonuses for wind-generated power in most countries that spring from an ambition to support the development of renewable energy sources and to reduce emissions that harm the environment and cause external costs to society. Some countries have a CO_2 tax reduction (Denmark), others have green certificates (Sweden, Great Britain, Italy), or offer a special long-term purchase price for the power, so-called feed-in-tariffs (FIT) (Germany, Great Britain, Spain).

Since rules, regulations, conditions, taxes and market situations differ, a specific analysis has to be made for each country. Since rules and regulations are changed and the future market prices are not known, even this is a very complicated and uncertain task. National wind power associations, state energy authorities or ministries can provide this kind of country-specific information on rules, tariffs and so on.

Calculation of economic result

To calculate the economic result, an assumption about a price per kWh for the next 20 years has to be made. This assumption has to be based on the facts that are known when the calculation is done. Since this calculation will be the basis for a decision about the investment, it should be supplemented with a *sensitivity analysis*, with a *worst-case scenario*, and also a *best-case scenario*. By doing this the economic risk can be evaluated. The higher the estimated risk, the more expensive it is to borrow money for the project – higher risk implies a higher interest rate.

The economic result, i.e. the annual profit, is calculated in this way:

$$P_a = I_a - C_a - OM_a$$

where:
P_a = annual profit
I_a = annual income
C_a = annual cost of capital
OM_a = annual cost for operation and maintenance.

The investment has been financed by a loan, which gives an annual cost for capital during the years when the loan has to be paid back to the bank, including the interest.

Cost of capital

The annual cost for capital is calculated by the annuity method. The annuity is the sum of amortization (pay back of loan) and interest where the sum of the amortization plus interest will be constant, i.e. the same sum each year (see Table 6.3). The annual capital cost C_a is calculated by the so-called annuity formula:

$$Ca = a \times C_i$$

where:
a = annuity
C_i = investment cost.

Present value and IRR

Another method to calculate the economic result for an investment is the *present value* method, which also is called the *discounting* method. With this method the size of an annual income or expense that will occur for a specific number of years is given the value at a specific time, usually the day when the turbine starts to operate. If the present value of the revenues is larger than the present value of the investment and expenses, the investment will be profitable.

The IRR – internal rate of return – is the rate of return when the present value is set to zero. It gives a measure of the annualized effective return rate which can be earned on the invested capital – the yield on the investment.

Payback time

A third method to evaluate the economic preconditions for an investment in a wind power project is the payback method. It is used to calculate how long it will take to get back the invested money. The simple payback time is calculated with this formula:

$$T = \frac{\text{Investment}}{\text{Annual net income}}$$

Levelized cost of energy

The actual cost to produce 1 kWh with wind power is an interesting figure. This energy cost is equal to the annual capital cost plus the annual operation and maintenance cost, divided by the annual production in kWh; this is called the *levelized cost of energy*.

$$E_{\text{cost}} = \frac{C_a + OM_a}{\text{kWh/year}}$$

Cashflow analysis

A cashflow analysis is a good method to calculate the economic result year by year. It shows the cashflow during the economic lifetime of the wind power plant, and can be performed with a spreadsheet software. Information on calculated power production, power price, certificates and other bonuses,

Table 6.3 Annuity factors

	Payback time			
Interest rate	*5 years*	*10 years*	*15 years*	*20 years*
5 %	0.2310	0.1295	0.0963	0.0802
6 %	0.2374	0.1359	0.1030	0.0872
7 %	0.2439	0.1424	0.1098	0.0944
8 %	0.2505	0.1490	0.1168	0.1019
9 %	0.2571	0.1558	0.1241	0.1095
10 %	0.2638	0.1627	0.1315	0.1175

loans, interest rates, and other factors that have an impact on the calculations are entered into the spreadsheet. Expected inflation rates and increases in power purchase prices can be entered, as well as rising costs for maintenance. The program then calculates the outcome year-by-year. The result can be presented in diagrams and tables showing the cashflow: annual revenues, capital costs, maintenance costs and remaining surplus. The interest rate has a great impact on the economic result. The higher the interest rate is, the longer it will take to pay back the loans.

Risk assessment

The calculation of the economic result is based on several assumptions. The first is the calculated power production. The second is the power price. What happens if these assumptions are wrong? To find this out it is always wise to assess the risk with a sensitivity analysis.

A realistic scenario for the worst things that can happen, a worst-case scenario, is made. This could be that the total power price will drop by 15 per cent and that the power production will be 10 per cent less than calculated (because of errors in the calculations, climate change, grid failures or other reasons). These figures are then used in the same calculations that have already been made. A similar best-case scenario should be made as well. The best case could be that the total power price will rise by 20 per cent and power production will be 10 per cent higher than calculated (because of errors in the calculations, climate change or other reasons).

The worst risks for the power industry are unexpected increases in fuel prices. This can never be a problem for wind power since the fuel is free. There is, however, uncertainty about the energy content of the wind that should be considered. The economic calculation for the base case is based on the expected power production during a normal wind year. During a specific year in the 20–25 years that the wind power plant will operate, the energy content in the wind can vary considerably. Some of these years the wind power plants could produce 20 per cent less than average. The estimated annual production for the 20-year period can also be too optimistic (McLaughlin, 2009).

This uncertainty should also be included in the sensitivity analysis, and combined with the price uncertainty. The absolute worst case is that the price will drop by 15 per cent and that production is 15 per cent less than expected. The outcome of this case should be above the levelized cost of energy from the wind power plant. There is also a very best case, where prices as well as production will be higher than expected. There is a risk and an opportunity. To evaluate these is a matter for the investors and the owner/operator. To do that a well-founded risk-analysis is necessary. Such risk assessment can be made by the wind power software, for example the loss and uncertainty module in WindPRO.

Financing

Wind power plants have a high investment cost but very low operational cost since the wind which is used as fuel is free. In this respect wind power is similar to hydro power. This means that in most cases significant loans are necessary to finance wind power projects.

The project developer has to research loan and grant options, and find banks or other investors that offer interest rates and payback times that the project can afford. Most banks nowadays consider investments in wind power like any other investment and don't add any extra risk premiums. Up to 70–80 per cent can usually be financed this way with the wind power plant as the sole security. The remaining funds have to be supplied from the company's reserves or by using some other asset as security (land, real estate, etc.) or from equity. If the last option has to be used, the developer has to raise equity and negotiate and execute agreements with equity investors. Investors will require a higher return on investment than the banks.

How a wind power project is financed depends on what kind of owner it has. In many countries wind power plants are developed and owned by independent power producers (IPPs), and not by big utilities and power companies.

Big companies often have a good cash balance and reserves and, thus, the capital needed for the investment is available within the company. This is called corporate financing. The other financing principle is called project financing, and can be also utilized by large corporations (Redlinger, 2002). In this case the wind power project is treated as an independent economic entity. Small and medium-sized enterprises which have been formed with the sole purpose of owning and operating wind power plants usually have to take loans from a bank or other credit institution. A limited company will raise some or all the money for the investment from equity.

Documentation

A wind power project has to be well documented. Readable, accurate and reliable documents that describe the project are needed to inform the public, for the permission application, for financing and also for investors who are prepared to realize the project. Some of these documents are public, others are internal documents for the project development company.

Project description

The project developer makes a feasibility study, which is an internal document. The project has, however, also to be made public and for that a good *project description* has to be prepared. This is a document intended for non-professionals and should be understandable and also address the questions the public are expected to raise. It should describe the area, the number of wind turbines and their siting. Impacts on neighbours, sound emission, shadow flicker, access roads and new power lines should also be described. Finally the visual impact should be illustrated by photomontages.

An introduction should contain information and arguments for the necessity of developing renewable energy sources – climate change and mitigation, the positive impact at a global level. A presentation of the developing company which will own and operate the wind power plant when it is installed should also be included in the project description.

This report should be non-technical, have many maps and other illustrations, basic information on wind turbines, size and annual production. If there are no good writers in-house, it is a good idea to engage a professional writer to compile this public report. Having a good project description will make it possible to engage in a good public dialogue, and it will create goodwill for the project development company.

Environment impact assessment

The project description can also be used for the dialogue with authorities in the permission process. For larger projects there is also a requirement for a formal environment impact assessment (EIA). The requirements for an EIA may differ in different countries and for different projects (Donnelly, 2009). What the EIA should cover is decided in the dialogue with the authorities. This report should also start with a non-technical description, and the project description can be reused for this. This is, however, a much more comprehensive report, often with technical and scientific reports from investigations made on grid connection, access roads, bird life, etc.

An EIA should describe the impact of the wind power plant on the global, regional and local environment. Each impact (on the landscape, sound emissions, etc.) should be described on three levels: first the present situation, then the impacts (change, consequences) and finally precautions (measures that minimize impacts). The positive impact from wind power, by reducing greenhouse gas emissions, should always be included in the EIA.

The impact during different stages of development should then be described. How the building process will be organized: preparatory work, access roads, power lines, working areas for cranes, excavators, trucks, storage places, etc. The impact during operation should be described in detail: visual impact, sound propagation, shadow flicker, safety measures. Finally the restoration of the site should be detailed: how the wind turbines will be dismantled and the ground restored, and how this process will be financed.

In a complete EIA at least two different options for siting and/or design of the project and a so-called zero-option should be included. The different options and their environment impact should be described in a way that makes it possible to compare them and to assess which option will be the best for the environment. The zero-option should describe the consequences if the project is not be realized. It could describe how the electric power that would be produced by the wind power plants will be supplied (by coal or natural gas and the emissions that will be generated).

Measures that will be taken to prevent, reduce or compensate for impacts and other precautions should be presented in separate paragraphs. Measures to prevent damage during the building process should be described here. Finally a summary of the process of public consultation and dialogue with authorities that is the basis of the EIA-report should be included.

When the application for permission is submitted, the EIA report will be public. It will be scrutinized by experts, as well as by possible opponents. The EIA has to be of a high quality, and if there isn't in-house ability to prepare a good EIA, consultants can be engaged to prepare it and to undertake some of the investigations that are included in the EIA.

Economic reports

The economics of a wind power project are business matters, and the business calculations of the developer are internal documents. However, to be able to finance projects, banks, financial institutions and private investors will require very detailed information on the project's economic viability, and on risks. Potential investors will also conduct a due diligence process, and check that contracts with landowners and grid operator are legally correct, that all necessary permissions are granted, and that a PPA is signed on reasonable terms.

Wind data report

The annual energy production is the basis for the economic viability, and a wind data report is the core of the economic evaluation. The report that describes in detail how the wind resources have been estimated is, thus, the most important document. This document should describe in detail how the local wind climate has been researched. The requirements of banks on the methods used are very strict, at least for large projects. The methodology, measuring equipment, certificates for calibration of anemometers used, as well as the measurement period, collected wind data and how they have been transformed to long-term data has to be detailed. As a rule, banks and other investors will demand a third-party evaluation of this report from an independent international consultant. Therefore it is wise to consult potential financiers before measurements start and ask them to specify their requirements.

Economic prospect

To finance the project it is always necessary to have well-founded and convincing documentation of the project's economic prospects. The developer has to prepare a budget, cashflow, balance sheet and income statements, to prove the case to banks and equity investors. This can be done in a prospectus for potential equity investors, which describes the opportunities as well as the risks with an investment.

Real budget

None of the calculation methods described above are sufficient to work out a real budget for a wind power project. They are used to come to a conclusion on the economic feasibility of the project. To make a real budget, a much more detailed analysis, based on tenders, actual credit conditions from banks, the share of equity vs loans, etc., has to be made. Interest rates have a great impact on the economic viability and result, and these are influenced by the perceived risks and opportunities of the investment. The real budget, and the outcome, is an internal report.

Building a wind power plant

The building phase includes all activities from when work to prepare the site starts until the handover of the operational wind power plant to the buyer/owner.

For small and medium-sized wind power plants, the time from placing an order for wind turbines to delivery to the site can vary from a few months to a year or more. In the meantime, access roads, turbine foundations and substation, and the electrical infrastructure have to be built. Furthermore, installation is weather sensitive: installing the wind turbines during the rainy season is not a good idea, neither is doing it in windy conditions. Delays due to bad weather conditions will increase cost, but can be avoided by good planning.

The developer has to prepare the site and build the foundations. Wind turbine foundations of reinforced concrete have to harden for a month. The wind turbines are usually mounted and installed by the turbine supplier. Installing the transformer (if it is not integrated into the turbine), the internal grid and the connection to the public grid are tasks for the developer.

Wind turbines are large and heavy, so suitable site access for transport is required. The transport vehicles are also quite long, and may have problems with bends on the road and up and down hills. The supplier usually checks out the transport route, and to make it work some trees may have to be felled on sharp bends, telephone and power lines raised or temporarily removed. On-site storage and assembly work will require an open space at the base of each tower of approximately 50 × 70 metres. Heavy cranes will be required on site; a 2 MW turbine requires a crane with capacity to lift several hundred tons to hoist the tower, nacelle and rotor into position.

When the wind turbine components finally arrive, it does not take more than a day to mount it on site, if everything is well prepared and the weather allows. Once the wind turbines are installed, about two weeks are needed to complete the installation work, commission the system and connect the wind power plant to the grid. The building phase for a small or medium-sized wind power plant only lasts two to three months.

Selection of suppliers

When all necessary permissions are granted and financing arranged, it is time to find a supplier for the wind turbines and other equipment, contractors to build access roads, and other tasks that have to be done to complete the project. A review of technical specifications of wind turbines should be done to determine which model would be most suitable for the project in terms of capacity, size, price and availability.

In the procurement of wind turbines, tenders should be invited from several different suppliers. A tender enquiry document (TED) should include basic information about the project timeline and planning status, financial, technical and operational information, contractual issues, scope of supply, technical specifications, maintenance and repair conditions as well as agreements for warranty and insurance.

The tenders have to be evaluated to find the one that is most favourable. This is not always the supplier which offers the lowest price. The supplier's record, ability to offer good service and maintenance and other factors should also be considered. At this stage the supplier may ask for a down payment of 10–30 per cent for the wind turbines that are ordered, when the contract is signed.

The developer has to negotiate and execute a turbine purchase agreement and warranty, and place an order with the required deposit. To avoid problems with warranties the supplier should make a formal statement that the delivered wind turbines are fit for the site where they will be installed. During the building phase many different processes have to be coordinated, so a delivery schedule for the wind turbines should also be included in the turbine sales agreement.

Contracts

Contracts have also to be signed with landowners, the grid operator, and with a power company, utility or third party that will buy the power. Contracts for loans from banks and other financial institutions should be signed as well. Then there are contractors who have to be engaged to build access roads, pads for cranes and equipment, transformers, cable ditches, electrical work. Potential construction companies have to be investigated, cost estimates worked out and construction contracts signed. It is good practice to engage local companies and contractors to build access roads, foundations and other tasks, as this gives benefits to the local community. Since all of this work has to be coordinated, a detailed timetable has to be worked out for all these contracts.

Supervision and quality control

It is common practice to assign a number of 'checkpoints' as part of the quality control process during production of wind turbines and construction of a wind power plant. These checkpoints give the owner an opportunity to audit the progress and quality of work and to verify that the components conform to the specifications. Checkpoints are mostly planned immediately following a project milestone, like factory acceptance test of components ready for shipment, site acceptance test when components are delivered to the site and several inspections during building and installation.

Commissioning and transfer

When the building phase is completed, before the wind power plant is handed over, an overall inspection and commissioning is carried out. Representatives of the contractor, the owner and the grid operator perform these final commissioning inspections. The commissioning involves a comprehensive testing and monitoring plan. The main objective is to verify that the system is complete, installed correctly and that it works properly.

The procedure for commissioning is formulated in mutual cooperation, with all the parties involved. Often defects are discovered during the inspections, and these defects are listed. The parties decide what actions should be done to clear this list, and sometimes a second inspection is required. When the commissioning has been approved, final project payment is made. The wind power plant is then transferred to the client, the owner/operator.

Operation

From the date of handover the owner is responsible for the daily operation of the wind power plant. Now warranty and maintenance contracts come into force. The terms of warranties can vary for different suppliers. They can be valid from two and up to twelve years from the handover, and include repairs and modifications. Some warranties also cover technical availability of individual wind turbines and 95 per cent of the certified power curve. Warranties do not cover actual performance since the available wind energy at the site cannot be accurately predicted; the winds can vary quite a lot in different years.

If the availability or performance is less than the warranted value, the supplier has to settle the difference. Besides the warranty, insurance cover is required for third-party liability, machine breakdown, lightning strikes, fire, vandalism, and insurance for business interruption during unproductive days following a breakdown. Financial institutions require adequate insurance as a precondition for loans. Machine insurance is not necessary until the warranty has expired.

The supplier insures against damage during transport and installation of the wind turbines. After commissioning and transfer of a wind power plant, the owner takes over all risks, excluding those covered by the warranty.

Maintenance

When the wind turbines have been installed, connected and started to feed electric power to the grid, they will operate unattended. The owner or those in charge of operations can monitor them from from an office desk, since the wind turbines' control system is connected to the operator's PC. Simple operational disruptions can be attended to from a distance, and the turbine can be restarted from the PC.

When more serious disruptions occur, the operator has to visit the wind power plant to rectify the fault before the wind turbine can be restarted. When a fault occurs, the operator will get notification by cell phone or PC.

Modern wind turbines require regular maintenance servicing twice a year. The wind turbine manufacturer can do this servicing and this is often included in the warranty for two years or longer. After that a contract for regular servicing should be negotiated with the manufacturer or an independent servicing company. It is also recommended that the owner/operator create a contingency fund and set aside funds for major repairs that may have to be done after the warranty has expired.

For a wind turbine in the MW-class, planned preventive maintenance takes two technicians two to three working days. They will inspect and test the control and safety devices, repair small defects, and replace or replenish gearbox lubrication. Oil samples are taken from the gearbox at regular intervals and are analysed. Filters are replaced and gears are inspected for damage.

The need to repair wind turbines varies widely between different wind power plants, but three to four mechanical or electrical faults a year that require a visit by a service engineer is a likely average. The downtime for each failure can last for one to four days. However, downtime can be much longer if spare parts are hard to locate. This is often the case with older wind turbines that are no longer manufactured.

The amount of power the wind turbines produce is registered on a meter that the grid operator will monitor. Settlements are usually made once a month, when the owner gets paid for the power that has been delivered to the grid in the preceding month. Rules for how this is done should be included in the power purchase agreement.

Condition monitoring

To reduce downtime due to technical faults, systems for condition monitoring have been introduced. A modern wind turbine has a comprehensive condition monitoring system from the start, which checks temperatures, rotational speeds, voltage and many other parameters. When a parameter falls outside specified values, the turbine is stopped and notification is sent to the operator.

Condition monitoring systems have some additional sensors, and software that analyses vibrations, frequencies, etc., and sends notification when some component might be worn out are indicated. The idea is to fix a problem before it results in a shutdown, and avoid the downtime by repairing or changing components before they cause the turbine to stop. For wind power plants in remote locations and offshore, where it takes a long time to carry out repairs, this will reduce costs.

Performance monitoring

Performance monitoring means that performance is continually checked in relation to the wind conditions, i.e. that the turbine at all times produces power according to its power curve. The reasons for a turbine to under-perform may be caused by other factors than faults in components, and these will not be identified by the control system or the condition monitoring system. One example is that the yaw system sets the turbine rotor disc a few degrees off from the perpendicular wind direction, another is a fault in the equipment that measures the power fed to the grid.

Decommissioning

After some 20–25 years of operation, when a wind turbine is worn out, it should be dismantled. Most of the parts can be recycled as scrap metal. The only components that can't be recycled today are the rotor blades, but there are efforts to find methods to achieve that as well. The scrap value of a turbine is about the same as the cost to dismantle it. The foundation of reinforced concrete, built below ground level, can be left behind, if it does not affect ground conditions in a negative way. Otherwise it can be removed and reused as filling material for roads or buildings.

When the wind power plant has been dismantled, it leaves no traces behind. However, if it is a good site, a new generation of wind turbines will be installed. This repowering process has already started in Denmark, Germany and Sweden, and will increase the installed capacity since new turbines have higher capacity than those that are replaced.

Summary

Designing and implementing wind power projects requires many different skills. The most important task is identifying good sites for wind power plants where the energy content as well as the quality of the wind (i.e. low turbulence) is high. From the power system's point of view it is, however, an advantage that wind power plants are geographically well distributed, so that the variation of the output is smoothed out in the system, which makes it possible to have a high wind power penetration. The wind resource is, thus, not the only criterion and since wind turbines can be tailored to suit the wind conditions at specific sites, it is possible to utilize wind power in many different regions.

The sites chosen for wind power development should also be uncontroversial. Sites where there are obvious conflicts with opposing interests should be avoided. Support from public opinion is important, as this can be earned with an open public dialogue and good information about how a project develops. It takes weeks and months to create such public trust, but it can be lost in a moment. So it is always worth the effort to spend reasonable time and resources keeping the public informed, and to consider the opinions that are put forward.

The size of wind turbines as well as wind power plants seems to be ever-increasing. Still there is no reason to focus just on larger turbines and wind power plants. Size should be well adapted to local conditions, loads and the capacity of the grid. There are advantages with distributed generation, as well as distributed ownership, and hopefully manufacturers will continue to manufacture turbines of different sizes.

The configuration of wind power plants, the micro-siting, is a very important aspect. Much can be gained by spending time and effort on this to avoid cramming a wind power area with too many wind turbines, which reduce park efficiency and also likely increase maintenance costs.

The methods of measuring the wind and calculating annual energy production are getting better and make it possible to increase the accuracy of the estimated power production. Getting this right is important; when production does not live up to expectations, investors will become disillusioned and it reduces their confidence in wind power development.

The technical availability of wind turbines is high, and with increased use of systems for condition and performance monitoring, it will become even higher. For good performance and cost-efficient operation of wind power during their technical lifetime, it is necessary to have good management of the operation and maintenance as well, which will reduce downtime and also increase the lifetime of the wind power plants.

References

Dahlberg, J-Å. (2009) *Assessment of the Lillgrund Windfarm: Power Performance, Wake Effects*. Stockholm: Vattenfall/Swedish Energy Agency.

Donelly, A., Dalal-Clayton, B. and Hughes, R. (2009) *A Directory of Impact Assessment Guidelines*. 2nd edn. Nottingham: International Institute for Environment and Development.

Earnest, J. and Wizelius, T. (2011) *Wind Power Plants and Project Development*. New Delhi: PHI Learning.

EWEA (2009) *Wind Energy: The Facts*. Brussels: EWEA

Krohn, S., Morthorst, P-E. and Awerbuch, S. (2009) *The Economics of Wind Energy*. Brussels: EWEA.

McLaughlin, D., Clive, P. and McKenzie, J. (2009) 'Staying ahead of the wind power curve', *Renewable Energy World*. Accessed 18 November 2014 at http://www.renewableenergyworld.com/rea/news/article/2010/04/staying-ahead-of-the-wind-power-curve

Milborrow, D. (2012) 'Wind energy economics' in A. Sayigh (ed.) *Comprehensive Renewable Energy*. Oxford: Elsevier.

Rathmann, O., Barthelmie, R. and Frandsen, S. (2006) *Turbine Wake Model for Wind Resource Software*. Roskilde: Risoe National Laboratory.

Redlinger, R.Y., Dannemand Andersen, O. and Morthorst, P. (2002) *Wind Energy in the 21st Century*. Basingstoke: Palgrave.

Schou Nielsen, B. and Stiesdal, H. (2004) 'Trees, forests and wind turbines: a manufacturer's view', presented at Trees Workshop. British Wind Energy Association, Glasgow, March.

van de Wekken, T. (2008) 'Doing it right: the four seasons of wind farm development', *Renewable Energy World*. Accessed 18 November 2014 at http://www.renewableenergyworld.com/rea/news/article/2008/05/doing-it-right-the-four-seasons-of-wind-farm-development-52021

Wizelius, T. (2012) 'Design and implementation of a wind power project' in A. Sayigh (ed.), *Comprehensive Renewable Energy*. Oxford: Elsevier.

Websites

RETScreen. Available at www.retscreen.net
WAsP. Available at www.wasp.dk
Wind Atlas. www.windatlas.dk
WindFarm – ReSoft. Available at www.resoft.co.uk
WindFarmer. Available at www.gl-garradhassan.com
WindPRO. Module description. Available at www.emd.dk.

Index